Agile Adoption and Transformation

Agile Adoption and Transformation

Principles, Challenges, and Pitfalls

Brian Will

SECOND EDITION
For permission requests, write to the publisher, addressed "Attention: Permissions Coordinator," at info@paroxys.com.
For U.S. trade bookstores and wholesalers order,
please contact the publisher via the contact above.

Cover design by Chris Babental, www.babental.com
Interior layout and design by www.writingnights.org
Book preparation by Chad Robertson

ISBN: 978-1-7373374-3-0

Library of Congress Cataloging-in-publication Data:

Names: Will, Brian, author
Title: Agile Adoption and Transformation –
Principles, Challenges, and Pitfalls 2020 Edition / Brian Will
Description: Paroxys Press, Reston, Virginia, 2021
Identifiers: ISBN 978-1-7373374-3-0 (Perfect bound) |
ISBN 978-1-7373374-2-3 (eBook)
Subjects: | Non-Fiction | Agile | Computer software | Business process |
Classification: Pending
LC record pending

Paroxys Press

Printed in the United States of America.
Printed on acid-free paper.

24 23 22 21 8 7 6 5 4 3 2

To Sayuri – for giving me all your unconditional love.
To Ian – for showing me all the possibilities of an open mind.
To mom, dad, and my brother – I miss you.

CONTENTS

PART II

AGILE ADOPTION CHALLENGES

PART III
AGILE ADOPTION PITFALLS

LIST OF FIGURES

List of Tables

ACKNOWLEDGEMENTS

We all stand on the shoulders of giants, and without them I could not have had such a fun career. It has been a blast, in large part due to the incredibly smart, dedicated, funny, and outright brilliant people I have had the honor to work with.

Some people that I am deeply appreciative of are:

Armand Aghabegian, who gave me my first job in the US after I relocated from Germany. Thanks for taking a chance!

Kent Beck, who explained to me that automated unit testing really required a standardized framework in order to scale. Thanks for opening my eyes!

Josh Sharfman, who showed me that you can make hard business decisions and keep your humanity at the same time. Thanks for giving me hope!

Roberto Edwards, who always offered completely unvarnished feedback. Thanks for always having my back!

Sayuri Bryant, who taught me the value of truly listening to the voice of the customer. Thanks for never backing off!

All the great software engineers, test engineers, analysts, managers, and executives that put up with me. Without you, I would not have learned anything!

Thank you all.

INTRODUCTION

Why would you want to read this book? Because you are either in charge of, or part of, a team trying to move from an established process methodology to Agile. You are supposed to help your team, group, department, division, or company become more Agile. Congratulations!

There are thousands of books, articles, and blogs out there talking about one or the other Agile / Scrum subject:

- Books that explain the basic Agile / Scrum concepts.
- Books focused on specific Scrum roles: How to be a good scrum master, product owner, scrum team member, or Agile coach.
- Books that are focused on more narrow subjects: Agile thinking, how to write good user stories, how to manage the prod-

uct backlog.

- Books how to develop user interfaces in an Agile matter – Agile UX.
- Books that help you obtain certifications.
- Books that talk about various Agile flavors, such as eXtreme Programming, or how to scale Agile processes in an enterprise.

It is a testament to our industry and the success of the Agile Manifesto[1], Scrum, and various Agile pioneers that we have all this variety and all this information available.

But, how do you tie it all together? How can you transform your organization and truly become Agile, become more competitive, shorten time-to-market windows, and compete with the best? After all, you have a day job and you are supposed to keep things going while adopting a new process and improving things.

Also, the challenges of Agile Adoption and Transformation are vastly different, depending on the environment.

A lean and mean garage startup might have totally different challenges, and advantages, as compared with a well-funded startup – think about the freedom of being broke with no way to go but up versus the pressure of managing investors and an ever-increasing burn rate!

The issues faced by a medium-sized software company might be

---

[1] http://www.agilemanifesto.org/

radically different from the issues faced by a large multi-national behemoth producing industrial machinery or an international consulting firm selling mostly professional services.

A market leader of the past, on a slow decline due to changing technologies and changing markets, might embark on an Agile transformation effort in order to stay competitive or hold on to their market position.

This book focuses on the main issues that most projects and companies are facing when they venture down the path of an Agile adoption and transformation effort:

- What are the main *principles* of Agile / Scrum?
- What are the main *challenges* you will be facing?
- What are common *pitfalls* and how can you avoid them?

This book is ***not*** about teaching you every little detail of Agile / Scrum – there are many good books that do that – but rather providing a brief overview of the Agile / Scrum principles and then addressing all the real world challenges you might face.

The book is organized into three main sections:

- **Part I** lays out the basic principles of Agile / Scrum – just in case you have not read other books about it if you need an executive summary.
- **Part II** addresses the challenges that are common in all kinds of organizations.

- **Part III** focuses on the pitfalls of Agile adoption. What are the traps that most organizations fall into when trying to transform their organization – and how do you avoid them?

Last but not least, why refer to Agile / Scrum? Why not another Agile methodology, such as eXtreme Programming or Kanban?

For two main reasons:

1. Scrum is not specific to software development and can be applied to all kinds of industries.
2. Scrum is by far the most popular Agile methodology in the market today.

CONVENTIONS USED THROUGHOUT THE BOOK

Typography Clues

Whenever an important term or concept is introduced, it appears in **bold** type.

Direct quotes from the giants of the field, as well as points being emphasized, appear in *italic* type.

ICONS USED TO HIGHLIGHT IMPORTANT POINTS

The Lightbulb

 Something noteworthy! Pay attention to these lightbulbs as they signify important issues to understand.

The Bomb

 Something bad! Literally a bombshell, pitfall, common problem, or issue that companies are facing and struggling with.

Part I

Agile and Scrum Principles

Chapter 1

Why Being Agile is Easy - and so Hard!!

Agile / Scrum processes are easy to understand. Any decent Agile coach and trainer can explain the Scrum process to complete novices in less than 30 minutes.

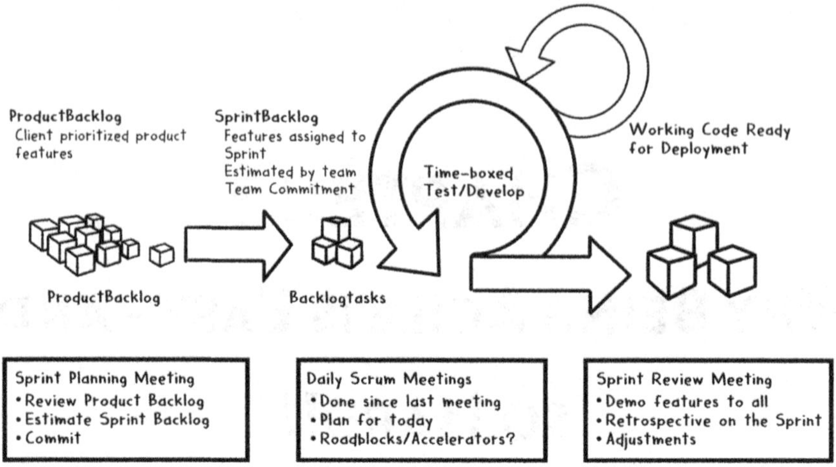

Figure 1 - Agile / Scrum Process

Admittedly, 30 minutes is not enough time to understand it in depth. However, the fact that the basics can be explained in that short a time is testimony to the simplicity and elegance of the process.

Anybody having observed a delivery-focused, productive, and effective Scrum team knows that there is "something else." Yes, there is the well understood process, the artifacts like the backlogs, or the ceremonies like the planning meetings, the daily standups, or the retrospectives. But that is not what makes a good team. There is something intangible that sets good teams apart from the rest of the pack.

So, why is being truly agile so hard? Why are many teams, despite following Agile / Scrum process frameworks to the letter, not producing the desired results? Why are so many Agile / Scrum teams perceived to not be agile? Why, after an Agile adoption process, are many teams delivering just as slowly as they did under their old waterfall process?

Here is what has been observed over the years – and it is a completely subjective list!

1. Old management structures stayed in place (meaning there was no organizational realignment).
2. Old processes were not adjusted (meaning many Agile / Scrum teams are forced to deal with two overlapping process models).
3. The team was not empowered to focus on execution (meaning decision making is still top-down, resulting in long decision making cycles typical in a command and control structure).
4. The team is focused on long discussions and plausible deniability instead of focused on getting things done and delivered.
5. The team does not understand the product / project goal or success criteria.

Easy ways to spot a team that is truly agile:

- Does the team regularly produce value for their stakeholders? Unless value is delivered all the time, with every sprint, the team is failing.
- Does the team validate their work to the best of their ability? Value needs to be delivered in working order. It makes no sense for a team to deliver a car where the wheels come off immediately upon trying it.

- Are stakeholders actively involved? Teams cannot operate in a vacuum and need stakeholder feedback; if teams are isolated, they usually fail.
- Is the team self-organizing? Teams needs to be able to adjust by themselves without management supervision. Who should do what? Who can help out?
- Does the team strive to improve its process? Teams need to have the ability to improve their own process. If process improvement stalls, productivity stalls.

Simply following an Agile process will not make you more agile. You really need to buy into the self-organizing team idea in order to be able to tap people's full potential.

You need to unburden the Agile team from other processes. Otherwise you simply ask them to do double the work.

You need to enable the team to fail, to learn, to self-correct, and to self-organize – all things that many organizations do not perceive as acceptable.

Teams that follow a traditional waterfall model ***can be*** incredibly agile, not because the waterfall process supports agility, but because the team has been empowered to be self-organizing, they self-corrected, they helped each other, etc. – in short, it is following Agile principles.

Agility is not a matter of the process that you follow, but a matter of enabling the intangibles to work their magic.

Don't do Agile, be agile!

Chapter 2
Agile and Scrum does not solve your problems

Software development blogs and the internet are abuzz about Agile / Scrum and other agile variations of the SDLC. Vendors of all colors and sizes are hyping Agile methodologies as the way to better and faster software delivery.

We know we have arrived at a place of complete buzzword penetration when industry representatives other than Silicon Valley software developers are talking about the need for "being Agile." President Obama's press secretary once mentioned the need for government to become "more Agile" and respond quicker to emergencies such as the Zika virus.

Everybody from the front line tech support representative to the company's CEO seems to talk about "Agile," "being Agile," or "becoming more Agile."

Of course, we are seeing unprecedented acceleration of delivery processes, yearly updates of major hardware and software platforms (think iPhone/iOS), almost constantly frictionless and transparent app updates on our phones, regular updates to our antivirus software, and public criticism toward companies that do not update product and services *all the time.* Social media hype cycles accelerate information flow to instant blips on a continuous stream of data floating by. There is also a lot of white noise about Agile / Scrum.

The problem is that many companies are jumping to the conclusion that adopting an Agile methodology will solve all their woes – without reflecting on the shortcomings or deficiencies of their current development capabilities. You have to know where you are in order to decide where you want to go.

An Agile methodology will not solve all your woes.

Simple truth is that even for simple processes, you must crawl before you walk, walk before you run, and run really well before becoming a track and field star.

You cannot skip levels. You have to learn each step of the process before progressing to the next one.

As simple as this principle seems, companies frequently try to go from an ad hoc, chaotic, and dysfunctional software development or

IT process to an Agile development methodology, hoping to improve software delivery cycles, quality, and customer satisfaction. This kind of silver bullet thinking will lead to great disappointments.

Malcolm Gladwell documented this principle of "crawl before you walk" in his book, **"Outliers,"**[2] where he makes the point that you need about 10,000 hours of experience in anything before you become really good at it. This principle applies equally to organizations. You cannot skip levels, from novice to expert, without putting in the time to hone your skills.

Maybe this is the reason that *some* companies that follow an old style waterfall process outperform their more aggressive, Agile-based, competitors? Maybe they just learned over the years how to do requirements analysis, project planning, coding, unit testing, continuous integration, and system testing in a way that allows them to reliably deliver their systems, whereas other companies that follow the latest Agile approaches are struggling through the learning process?

The author is a strong proponent of Agile delivery methods, but has healthy respect for companies that have made other methodologies work successfully for them. In short, it is preferred to have a well-organized development organization that uses a waterfall methodology over an ad hoc, crisis-driven, unorganized Agile development organization.

Agile / Scrum does not *automatically* equate to faster or better system delivery. Nokia had been dedicated to Agile development for

---

[2] https://www.amazon.com/Outliers-Story-Success-Malcolm-Gladwell/dp/0316017930/

years and long struggled to put competitive products out to market. We all know how that ended. So, Agile / Scrum does not equate to better and faster delivery or business success.

The point here is not to argue that waterfall is better than Agile / Scrum. Rather, the basic engineering principles still apply. Simply adopting an Agile methodology will not fix your problems if you have not mastered those engineering principles. You cannot skip levels.

An Agile methodology will not fix your problems if you have not already mastered software engineering principles.

Scrum is a product development framework and does not actually address software engineering practices at all. If you have not already mastered a software engineering practice, that should be part of your "becoming agile."

Having said that, if you have done your homework and all things being equal, Agile software development frameworks are superior to the older waterfall / iterative models.

Let's not forget the lessons we learned from excellent engineering companies (whatever development methodology they used) that died an unceremonious death because they missed the market. Good development methodologies and good engineering practices are a support function for achieving business success, not the other way around.

There are other factors that influence if and why a business can be successful that are beyond the scope of this chapter, but if you want food for thought you should watch this fascinating TED Talk by *Bill*

Gross: The single biggest reason why startups succeed[3]. Having reviewed success criteria for startups, he prominently points out that market timing seems to be of paramount importance. It's certainly something to keep in mind.

You can have great business success without Agile and without good engineering practices – and the business can fail completely despite using the best Agile and engineering practices. This is something Agile practitioners should keep in mind.

Agile / Scrum is not the magic bullet solution to all problems.

---

[3] https://youtu.be/bNpx7gpSqbY

CHAPTER 3
AGILE OVERVIEW

Although the terms "Agile" and "Scrum" have been around for more than 20 years and the Agile Manifesto[4] was signed by its signatories in 2001, it seems there is no shortage of Agile methodologies, frameworks, and flavors.

[4] http://www.agilemanifesto.org/

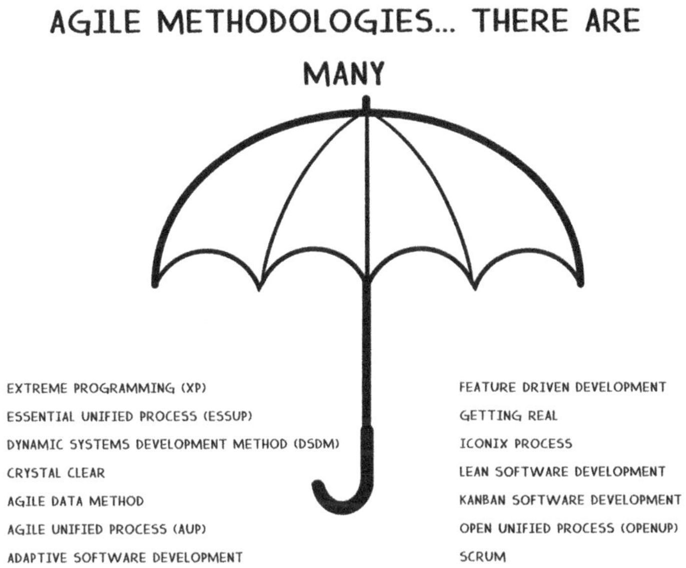

Figure 2 - The Growing Agile Umbrella

- Scrum
- Lean
- eXtreme Programming

The most popular flavor by far seems to be Scrum, especially considering that Scrum forms the basis for some of the other flavors. A search of Amazon in May 2017 yielded the following results:

- Scrum: 1,374 book results

- Lean Software Development: 289
- eXtreme Programming: 393

Considering the fact that Lean and eXtreme Programming are software development specific, this really leaves Scrum as the dominant general purpose Agile framework that can be applied to many kinds of environments, challenges, and development efforts.

For the purposes of discussing key principles, challenges, and pitfalls, this book will refer to Scrum. Although many of the issues discussed in this book are showcased referring to Scrum, there is nothing inherently Scrum-specific about them; many of the lessons can be applied to Lean or XP environments as well.

WHAT IS SCRUM?

The original meaning of the term "Scrum" goes back to Rugby, where it is defined as "*an ordered formation of players, used to restart play, in which the forwards of a team form up with arms interlocked and heads down, and push forward against a similar group from the opposing side. The ball is thrown into the scrum and the players try to gain possession of it by kicking it backward toward their own side.*"

The original thinkers must have had a Rugby player on the team! The Rugby-related meaning of the word has been lost, unless you play Rugby, and the for people who are not sports aficionados, the term "Scrum" seems like one of the many weird software-related acronyms. As such, you can see raging discussions online on whether we should spell the term as "scrum," "SCRUM," or "Scrum."

According to the Scrum Alliance®[5], "*Scrum is an Agile framework for completing complex projects. Scrum originally was formalized for software development projects, but it works well for any complex, innovative scope of work.*"

Please note that for the purpose of this book we will talk about Scrum when referring to the development framework and Agile when we talk about overarching principles or issues that generally are applicable to all Agile methodologies.

SCRUM IS SIMPLE

At its core, Scrum is composed of very few defined roles, artifacts, and events.

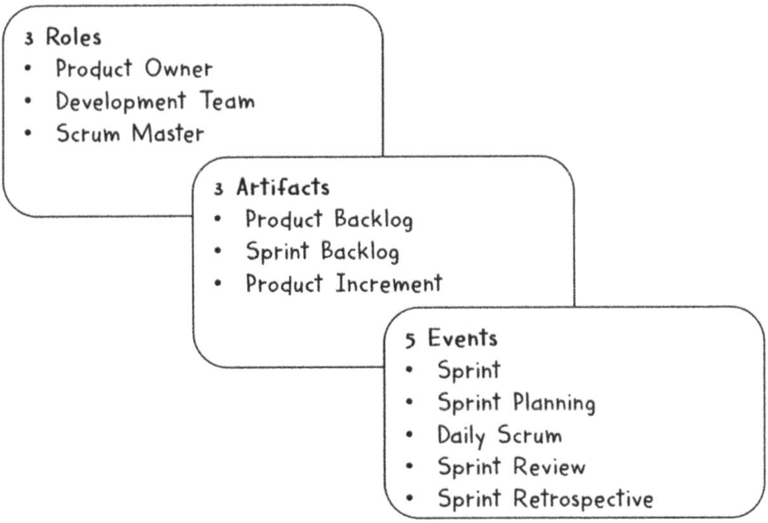

Figure 3 - Scrum Roles, Artifacts, and Events

[5] https://www.scrumalliance.org/

Because of the limited number of roles, artifacts, and events, Scrum is often referred to as a "lightweight framework" as opposed to other "heavyweight frameworks" that define many more roles, artifacts, and events in order to accomplish their purpose.

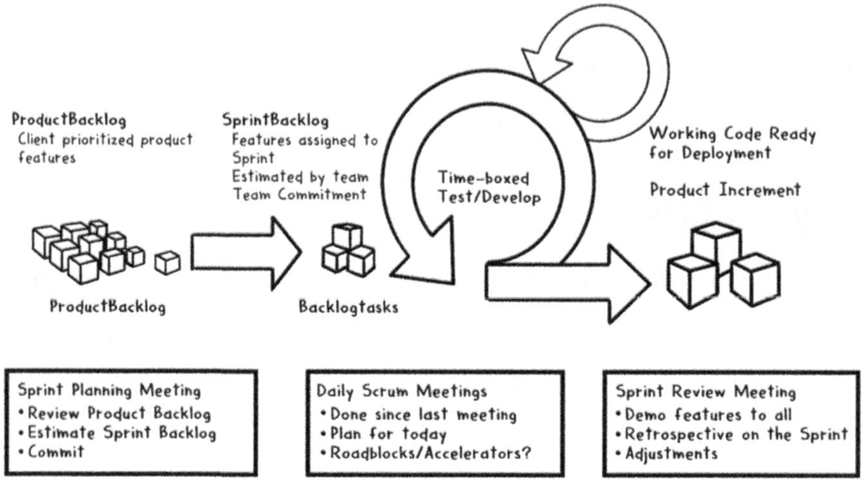

Figure 4 - Scrum Process Framework

This approach of being lightweight and easy to understand is very much in line with the old KISS principle – KISS is an acronym for "Keep it simple, stupid" as a design principle noted by the U.S. Navy in 1960. The KISS principle states that most systems work best if they are kept simple rather than made complicated; therefore, simplicity should be a key goal in design, and unnecessary complexity should be avoided.

SCRUM TEAMS ARE EFFICIENT

Similarly to the roles, artifacts, and events, the Scrum team structure is equally simple and straightforward to understand.

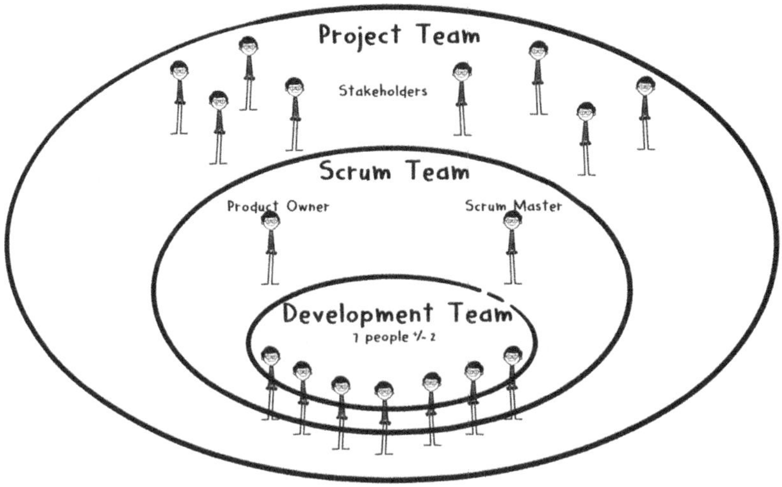

Figure 5 - Team Definitions in Scrum

AGILE HAS UNDERLYING VALUES

Underlying all this simplicity are basic principles that were published as the Agile Manifesto[6]:

We are uncovering better ways of developing software by doing it and helping others do it.

Through this work we have come to value:

Individuals and interactions over process and tools

Working software over comprehensive documentation

Customer collaboration over contract negotiation

---

[6] http://agilemanifesto.org/

Responding to change over following a plan

That is, while there is value in the items on the right, we value the items on the left more.

CORE AGILE VALUES

- Satisfy the customer through early and continuous delivery
- Welcome changing requirements, even late in development
- Deliver working software frequently
- Business people and developers work together daily
- Build projects around motivated individuals
- Convey information via face to face conversation
- Working software is the primary measure of progress
- Maintain constant pace indefinitely
- Give continuous attention to technical excellence
- Simplify: Maximizing the amount of work not done
- Teams self-organize
- Teams retrospect and tune their behavior

Reviewing the roles, artifacts, and events; the simplicity of the process; the composition of the (relatively) small team; and the underlying values, it becomes clear that the main focus areas are:

- Simplicity of Process
- Efficiency of Communication
- Delivery Focus

SCRUM IN THE CONTEXT OF THE REAL WORLD

The Scrum framework in and of itself works well for small, relatively isolated development efforts. Without getting into enterprise challenges (see later chapters), the reality is that many projects are done in existing organizations, with existing processes, active stakeholders, and already existing departments that might influence your process.

Hence, within the enterprise context, we are most likely faced with various phases of a product or project development effort that resembles something like this:

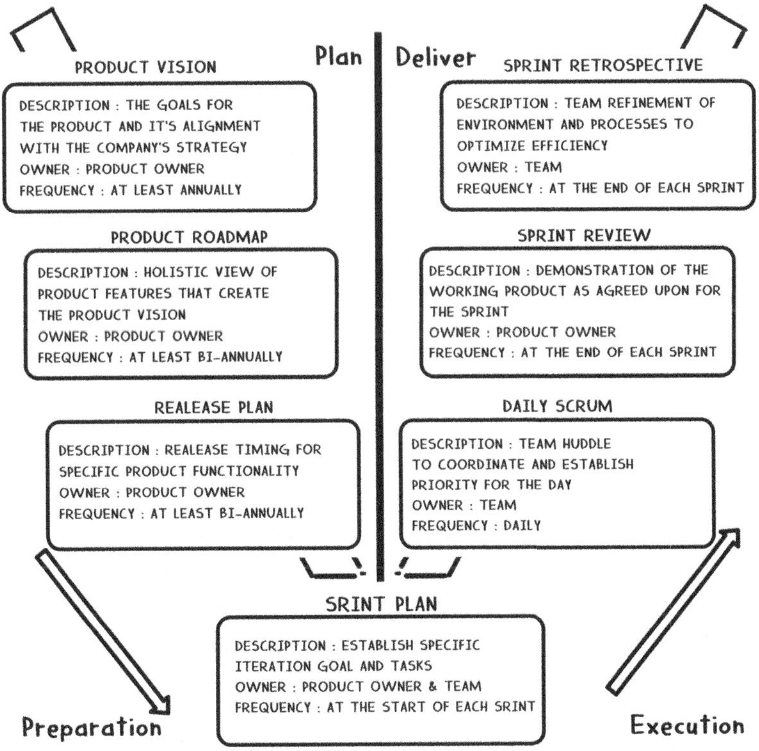

Figure 6 - Agile / Scrum Planning and Execution Phases

Although this can vary depending on your specific environment, almost all companies follow a process where planning and preparation precedes the execution and delivery of a product or project. The Pragmatic Marketing Framework™ specifically addresses strategy and planning phases that address product roadmaps, requirements management, and launch planning.

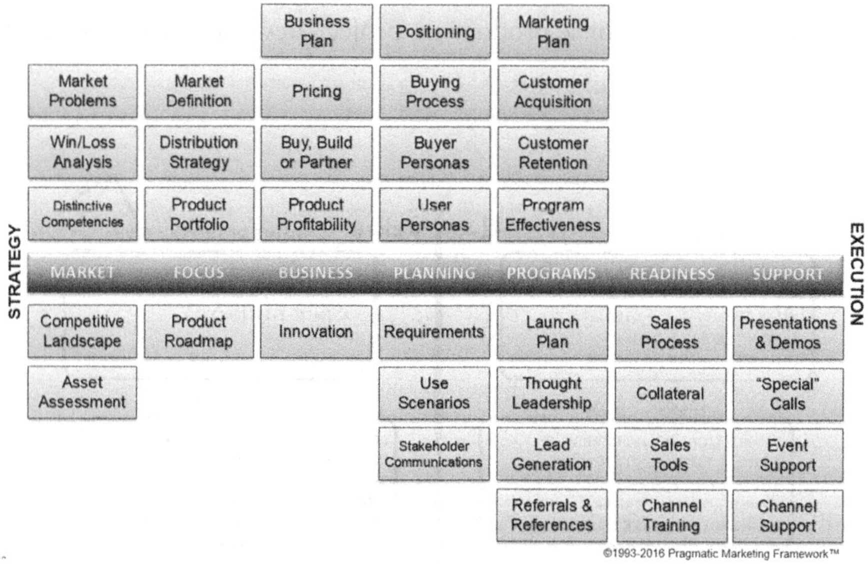

Figure 7 - The Pragmatic Marketing Framework™

So all product and project development efforts follow some kind of phased approach where the vision gets translated into a roadmap, which forms the basis for a high level plan, which gets refined into lower level plans, allowing you to deliver in an Agile / Scrum framework – but with a slightly expanded list of roles, artifacts, and events.

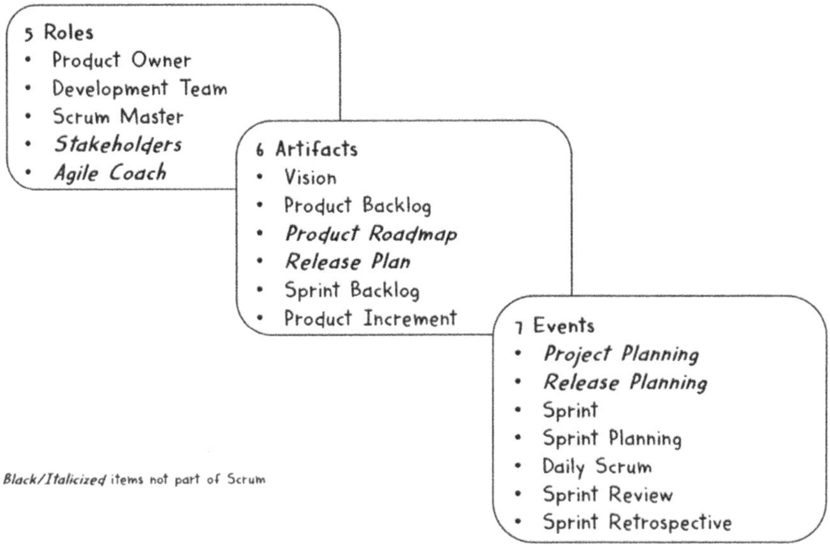

Figure 8 - Expanded Scrum Roles, Artifacts, and Events

Consequently, Scrum teams are faced with other influencing forces they need to consider. Depending on the specific environment that your development team will operate in, you might be faced with shared services groups, enterprise architecture teams, regulatory rules, etc. that will influence how the team can conduct its business.

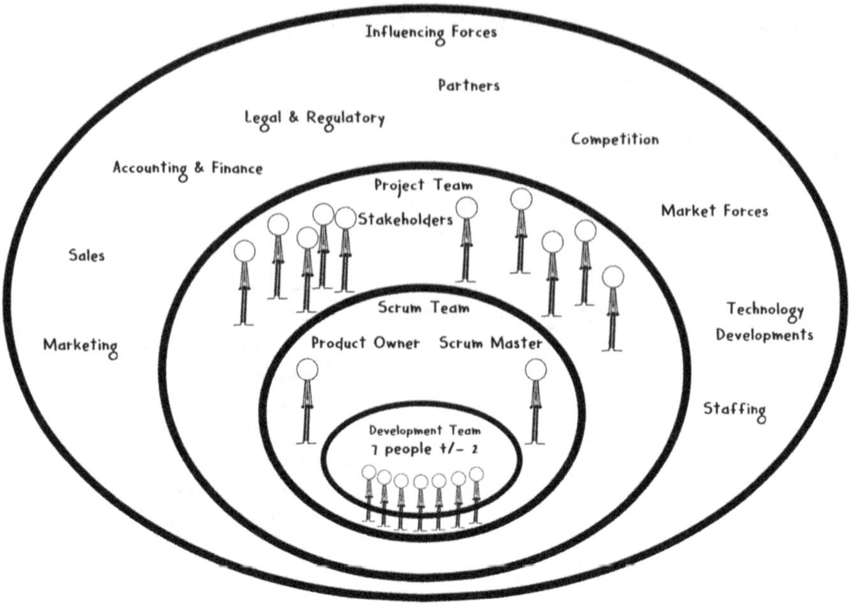

Figure 9 - Forces Influencing the Agile / Scrum Team

Enterprise development efforts often involve many teams working on one or several programs, which need to be coordinated both vertically and horizontally. Following is a simple Office Suite example showing the interaction and integration complexities across teams.

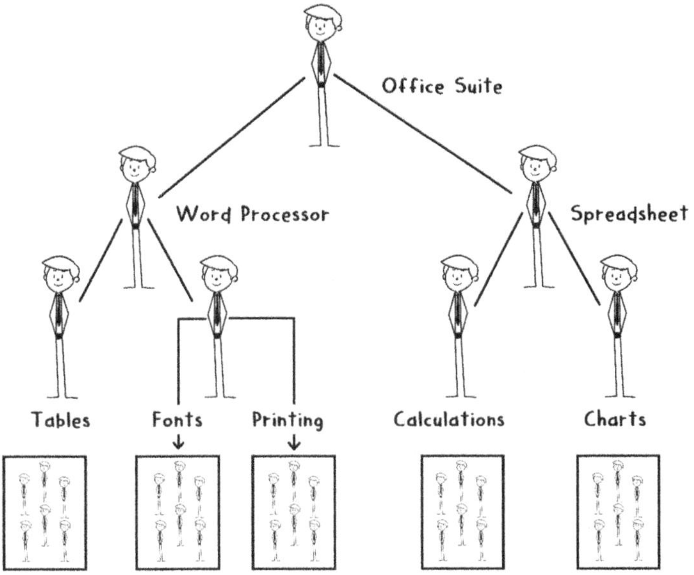

Figure 10 - Agile / Scrum Vertical and Horizontal Team Coordination, Functional Breakdown

Enablement of functionality, and integration of various systems, are the responsibility of the Scrum team above. Although conceptually easy to understand, this approach poses the biggest challenges when it comes to delivering complex systems. Hence the emergence of "enterprise scale" Agile frameworks, such as SAFe®.

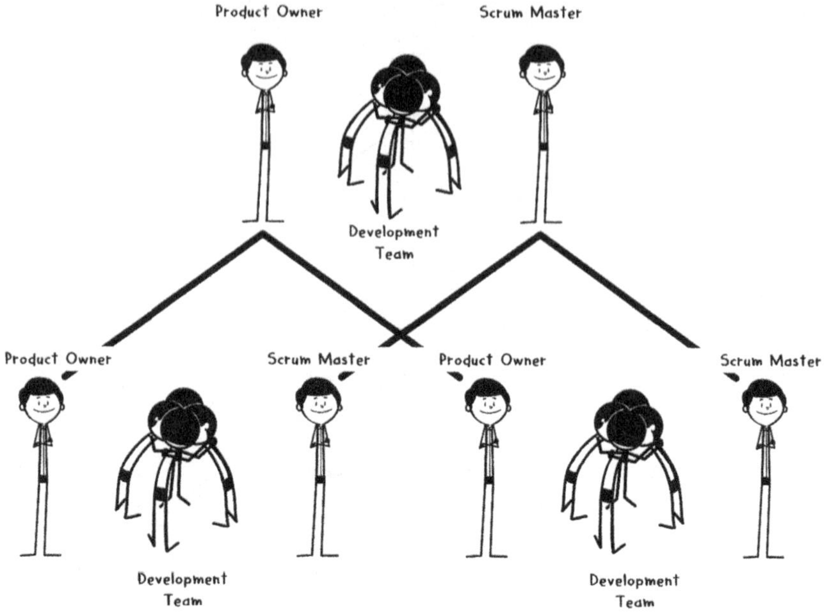

Figure 11 - Agile / Scrum Vertical and Horizontal Team Organization

RECAP OF SCRUM ROLES

- Development Team: The group of people who do the work of creating a product. Programmers, testers, designers, writers, and anyone else who has a hands-on role in product development are members of the development team.
- Product Owner: The person responsible for bridging the gap between the customer, business stakeholders, and the development team. The product owner is an expert on the product and the customer's needs and priorities. The product owner works with the development team daily to help clarify requirements. The product owner is sometimes called a customer representative.

- Scrum Master: The person responsible for supporting the development team, clearing organizational roadblocks, and keeping the Agile process consistent. A scrum master is sometimes called a project facilitator.
- Stakeholders: Anyone with an interest in the project. Stakeholders are not ultimately responsible for the product, but they provide input and are affected by the project's outcome. The group of stakeholders is diverse and can include people from different departments, or even different companies.
- Agile Coach: Someone who has experience implementing Agile projects and can share that experience with a project team. The Agile mentor can provide valuable feedback and advice to new project teams.

RECAP OF SCRUM ARTIFACTS

- **Product Vision Statement**: An elevator pitch, or a quick summary, to communicate how your product supports the company's or organization's strategies. The vision statement must articulate the goals for the product.
- **Product Backlog**: The full list of what is in the scope for your project, ordered by priority. Once you have your first requirement, you have a product backlog.
- **Product Roadmap:** The product roadmap is a high-level view of the product requirements, with a loose time frame for when you will develop those requirements.

- **Release Plan:** A high-level timetable for the release of working software.
- **Sprint Backlog:** The goal, user stories, and tasks associated with the current sprint.
- **Increment:** The working product functionality at the end of each sprint.

RECAP OF SCRUM EVENTS

- **Project Planning:** The initial planning for your project. Project planning includes creating a product vision statement and a product roadmap, and can take place in as little time as one day.
- **Release Planning:** Planning the next set of product features to release and identifying an imminent product launch date around which the team can mobilize. On Agile projects, you plan one release at a time.
- **Sprint:** A short cycle of development, in which the team creates potentially shippable product functionality. Sprints, sometimes called iterations, typically last between one and four weeks. Sprints can last as little as one day, but they should not be longer than four weeks. Sprints should remain the same length throughout the entire project.
- **Sprint Planning:** A meeting at the beginning of each sprint where the Scrum team commits to a sprint goal. They also identify the requirements that support this goal and will be

part of the sprint, and the individual tasks it will take to complete each requirement.

- **Daily Scrum:** A 15-minute meeting held each day in a sprint, where development team members state what they completed the day before, what they will complete on the current day, and whether they have any roadblocks.
- **Sprint Review:** A meeting at the end of each sprint, introduced by the product owner, where the development team demonstrates the working product functionality it completed during the sprint.
- **Sprint Retrospective:** A meeting at the end of each sprint where the Scrum team discusses what went well, what could change, and how to make any changes.

RECAP OF PRODUCT / PROJECT DEVELOPMENT PHASES

For the sake of simplicity, this section omits larger enterprise considerations, such as how functional groups such as IT, Product Marketing, Product Management, or Enterprise Architecture influence the product vision, the product roadmap, and the release plan. It focuses on the phases and core participants of a simple Scrum environment for clarification purposes.

Preparation and Planning

- The product owner identifies the **product vision.** The product vision is a definition of what your product is, how it will support your company or organization's strategy, and who will use the product. On longer projects, revisit the product vision at least once a year.

- The product owner creates a **product roadmap.** The product roadmap is a high-level view of the product requirements, with a loose time frame for when you will develop those requirements. Identifying product requirements and then prioritizing and roughly estimating the effort for those requirements are a large part of creating your product roadmap. On longer projects, revise the product roadmap at least twice a year.
- The product owner creates a **release plan.** The release plan identifies a high-level timetable for the release of working software. An Agile project will have many releases, with the highest-priority features launching first. A typical release includes three-to-five sprints. Create a release plan at the beginning of each release.
- The product owner, the scrum master, and the development team plan sprints, and start creating the product within those sprints. **Sprint planning** sessions take place at the start of each sprint, where the scrum team determines what requirements will be in the upcoming iteration.

Execution and Delivery

- During each sprint, the development team has daily meetings, sometimes referred to as **daily scrum** or **daily standup.** In the daily meeting, (no more than 15 minutes) each participant discusses what they completed yesterday, what they will work on today, and any roadblocks they have.

- At the end of each sprint, the team holds a **sprint review.** In the sprint review, the team demonstrates to the product stakeholders the working product created during the sprint.
- After sprint closure, the team holds a **sprint retrospective.** The sprint retrospective is a meeting where the team discusses how the sprint went and plans for improvements in the next sprint. Like the sprint review, a sprint retrospective occurs at the end of every sprint, before the next sprint starts.

CHAPTER 4
ENTERPRISE SCALING

Scrum defines the roles, events, and artifacts of its Agile product development framework, but the framework only addresses the work of one team.

Since the **Scrum Guide**[7] only addresses how one team works on a product, what happens when multiple teams need to work together on the same product? What about products that need to integrate? And how about a company that has decided to "Go Agile" and employs hundreds or thousands of software developers; how would all of these teams work together?

Working with multiple teams in an Agile environment requires a

---

[7] https://www.scrum.org/resources/scrum-guide

framework or processes beyond Scrum to address how teams interact and define additional roles, events, or artifacts.

In 2001 the first book on scrum, ***Agile Software Development with Scrum***[8] by Ken Schwaber and Mike Beedle, included a 15-page chapter on scrum for "multi-related projects" in which the term and concept of "scrum of scrums" is introduced. The same year, Jeff Sutherland wrote an IT Journal article about scrum and showed how "scrum of scrums" can be used to scale scrum "to any size."

Schwaber's 2003 ***Agile Project Management with Scrum***[9], also dedicates 10 pages to scaling Scrum. So, even though Scrum is a single-team framework, its use on larger multi-team projects has been practiced successfully from inception.

The "scrum of scrums" was the first additional event introduced to scale Scrum and help teams organize their work together on larger projects. The first Scrum book also introduced the idea of a "shared components" team as a new role or team type to help deliver and manage a set of shared components. These ideas are later formalized by Schwaber into Nexus, but the "scrum of scrums" remained the primary method of scaling Agile until 2016, when SAFe® became the most used process framework for Agile. (see ***State of Agile, Version One 2016***[10]) In addition to SAFe, LeSS, Nexus, DAD, and AUP have made inroads in the software development industry as scaling processes or frameworks.

---

8 https://www.amazon.com/Agile-Software-Development-Scrum/dp/0130676349/
9 https://www.amazon.com/Agile-Project-Management-Developer-Practices/dp/073561993X/
10 http://stateofagile.versionone.com/

SCRUM OF SCRUMS

The "scrum of scrums" became a practice born of necessity when Ken Schwaber and Jeff Sutherland both applied Scrum to large-scale projects. Schwaber was working with multiple scrum teams that had inter-dependencies. There were three teams working together, and he decided to have one member from two of the teams join the third team's daily standup, so that the visiting team members could understand the dependencies for their team and report them back in their own team's daily standup. This joint standup evolved into a separate meeting with representatives from each development team and became known as the "scrum of scrums."

In 1996, Jeff Sutherland created a team-based organization and used Scrum as the framework throughout the entire organization, from developers to executives. Managers ran a weekly "scrum of scrums," and executives ran one monthly.

In this model the "scrum of scrums" iterates as many times as it needs to get through all the layers of the organization, and in Sutherland's example, all the way up to the executives.

The scrum of scrums asks the same three questions as the daily standup, but adds a fourth cross-team question:

1. What has your team done since we last met to meet the sprint goal?
2. What will your team do today to help meet the sprint goal?
3. Do you see any impediment that prevents your team from meeting the sprint goal?

4. Do you see any impediment your development team will cause for other development teams?

This enables teams to cross-coordinate dependencies for the sprint and address the impact of those dependencies face-to-face on a regular basis. Some organizations run a development scrum of scrums daily and a management scrum of scrums weekly. The process is flexible enough to scale to any size and can be used creatively due to its flexible nature.

In addition to the scrum of scrums, there needs to be some kind of product owner coordination for the scaling effort to be successful, either through joint sprint planning and joint backlog management as suggested by Schwaber in ***Agile Project Management with Scrum***[11] or through some other cross-team planning coordination. If you can't plan the cross-team work together, it will be hard to successfully deliver it together!

LESS – LARGE SCALE SCRUM

LeSS (Large Scale Scrum) has been used to successfully create some of the largest Scrum-based organizations in the software industry. Craig Larman and Bas Vodde launched LeSS with the publishing of **Scaling Lean and Agile Development**[12] in 2008. They had successfully scaled Scrum and created team-based organizations with hundreds and even thousands of team members. They subsequently pub-

[11] https://www.amazon.com/Agile-Project-Management-Developer-Practices/dp/073561993X/
[12] https://www.amazon.com/Scaling-Lean-Agile-Development-Organizational/dp/0321480961/

lished two additional books on LeSS.

LeSS adds two supplemental events to scrum, a joint sprint planning event and an overall sprint retrospective.

Sprint planning is broken into two sessions: an initial joint session for all teams to plan the overall product sprint together, acknowledging cross-team dependencies; and a second session for individual teams to plan their work for the sprint.

There is one shared product backlog, but each individual team has its own sprint backlog. The sprint is planned, executed, and inspected at the same time for all teams, with a functional product increment ready at the end of the sprint.

For the retrospective, there are the usual scrum team-based retrospectives, but there is also an overall retrospective with members from all teams. This overall retrospective creates its own backlog for organizational improvement. These two events are the only addition to scrum events, but LeSS does address other aspects of product development, such as the overall organizational structure and the optimization of the business strategy.

There is one big impact with LeSS: the framework strongly suggests that the organizational structure be modified to a team-based organization with nothing but:

- feature development teams,
- a product owner team, and
- a head of the product group.

There is a fourth category called the "*undone*" department, which should not exist, but is the place where teams like QA, architecture, and business analysis live if the organization forces the retention of such groups in LeSS adoption.

The philosophy is that these functions should exist within the development team members for each team.

It is important to point out here what is missing: there is no project/program organization or project/program management office (PMO). Traditional command and control organizations cease to exist in a LeSS organization as their responsibilities are distributed among the feature teams and the product owner. Insisting on keeping such organizations will cause confusion and conflicts of responsibilities.

Also missing in LeSS are support groups or shared services teams such as configuration management, continuous integration support, or "quality and process."

LeSS organizations prefer to expand the existing team's responsibility to include this work over creating a more complex organization with specialized groups. Specialized support groups tend to 'own' their area, which leads to them becoming a bottleneck.

A complete change to the organizational structure of the product group is something that is considered essential to the success of LeSS. The reality, however, is that most corporate organizations might not be willing to make this change - consequently, LeSS is not for everyone.

The fundamental question is if the organization is looking for an evolution toward Agile or a revolution – LeSS very much assumes that the organizational structure will change.

LeSS has two flavors: Regular LeSS for 2 – 8 teams and LeSS Huge for more than 8 teams.

LeSS Huge organizes the teams into "requirements areas," and each area has its own product owner. There can be 2 - 8 teams in a requirements area, and the teams all follow the basic LeSS framework. These requirements areas form the structure for the entire product group.

Interestingly, LeSS does not follow the scrum of scrums model for cross-team communication, but rather suggests that team members attend the standups of teams they depend on and use other less formal forms of communication. In LeSS, the cross-team communication is the responsibility of the team members. Scrum masters do not act as project managers and do cross-team coordination, but rather remind the team members that they have the responsibility to communicate and see that it happens.

LeSS organizations do not include manager roles, but the manager role is accepted if an organization requires it. LeSS says that managers should change their function into that of "capability builders" who help see the overall picture, removing impediments, helping with problem solving, and growing individual's capabilities. LeSS encourages the teaching of problem solving, with the application of different problem solving techniques, such as the Five Why's, System Thinking, and the Three A's. LeSS asks executives to use systems thinking and to view their business as a system, agreeing on what system optimizations are their goals.

LeSS organizes by customer value and uses cross-functional feature teams to understand and build features, resulting in fewer handoffs and deeper understanding of customer usage. This more direct interaction creates much clearer understanding of customer needs. Feature teams are self-organizing, cross-functional, stable, and long-lived.

In LeSS, Communities of Practice (CoP) are created for the distribution of knowledge that is not related to product features. For example a "Scrum Master CoP" could be created for scrum masters to share information about their practice and how it relates to the organization. The same could be done for BDD/ATDD, Big Data, etc.

CoPs are not part of the organizational chart but do need to be supported with funds and resources to encourage knowledge sharing throughout the organization.

LeSS truly takes Scrum to the scaled level and adheres to Scrum and Agile principles. However, the radical change to organizational structure is one of the largest barriers to using LeSS, as many companies resist the re-engineering of the entire organizational structure.

SAFE® FOR LEAN ENTERPRISES

The Scaled Agile Framework (SAFe®) was created by Dean Leffingwell in 2011 and became the most popular scaling framework by 2016. The www.ScaledAgile.com site states that Dean Leffingwell's "*Scaled Agile Framework® is the world's most comprehensive model for adopting Agile across an enterprise.*"

The marketing literature for SAFe® also states that "70 of the Fortune 100 use SAFe," which shows that SAFe® fills a need for corporate IT.

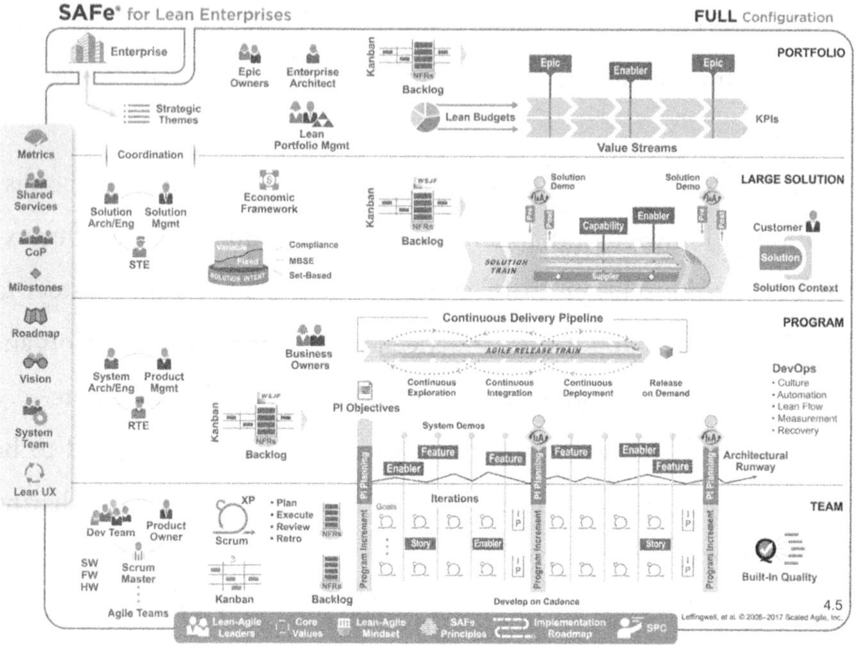

Figure 12 - SAFe® Full

There is likely no dispute that SAFe® is the most "comprehensive model" as stated by Scaled Agile, Inc. The framework takes a solution-oriented and encyclopedic approach, covering many of the concepts within Agile product development and even defining some.

However, many leading Agile practitioners will dispute that SAFe® is Agile, and say there should not be a comprehensive model. Many argue that SAFe® is positioned as a reformulation of older processes like RUP (Rational Unified Process).

The Agile 2014 conference had some lively online discussions around the topic, including one where Ken Schwaber, co-creator of Scrum, asked attendees to "*Welcome the SAFe people (even though they are not agile), because without a RUP conference they have nowhere to go.*"

So why is there so much controversy?

Many Agile practitioners prefer working with "just enough" in all aspects of product development. Scrum is the perfect example. It is a wide open framework, just a shell to be filed with Agile engineering practices and supported by additional product definition practices.

This leaves a lot of freedom for teams and organizations to implement whatever engineering and product definition practices work well for their organization.

SAFe®, on the other hand, offers up solutions "across the enterprise," literally providing a diagram which supports mapping traditional corporate IT roles into an Agile framework. SAFe® uses the term enterprise repeatedly and ties its framework to existing corporate IT organizational structures. This allows organizations to apply Agile concepts without radically modifying their structures, which is a much safer approach than LeSS, which suggests changing your organizational structure to match the Agile mentality. Simply put, how many PMO Directors want to hear "*We don't need a PMO*"?

SAFe® 4.0 was released in July 2016, and with the advent of SAFe® 4.5 in June 2017, the overall name was changed to SAFe for Lean Enterprises. There are now four flavors: Full SAFe, Large Solution SAFe, Portfolio SAFe, and Essential SAFe.

Full SAFe is a massive solution and starts at the top of the enterprise with business epics. The other three flavors are subsets of the Full SAFe.

ENTERPRISE SCALING – WHAT TO DO

As mentioned at the beginning of the book, it is a testament to the success of Agile thinking and the Scrum process that we are blessed with so many enterprise scaling flavors and variations.

It does represent surprising similarities to the "methodology wars" that raged within the object oriented programming community of the late 1980s and 1990s.

Personally, the author does not support one scaling framework over the other because "enterprise scaling" very much depends on several important factors, such as:

- Age of the organization – is the company a startup trying to scale or a 50-year-old company trying to rejuvenate its engineering approaches?
- Industry of the organization – is the company a pure software play? Hardware / software play? Manufacturer of entertainment electronics costing $15 / unit or producer of industrial machinery costing $2M / unit?
- Existing methodologies / processes – does the company follow an existing, established, process?
- Existing organizational structures – does the company have clearly defined organizational structures that exist to facilitate

product delivery, such as Product Management, Project Management/PMO, Development, Test Engineering, Production Line, etc.? Or are you still discovering what the organizational structure should be?

- Size of the organization – are you faced with an organization that has 500 or 50,000 employees?
- Locations of the organization – single location, geographically distributed across the US, or worldwide?
- Satisfaction with existing processes – is the company satisfied with how the existing processes work? Time to market delivery speeds? Revenue targets and profit margins? Customer satisfaction with products and / or services offered?
- What goals is the company trying to accomplish by adopting an enterprise scaling framework? Reduced time to market windows? Higher productivity? Competitive advantage?
- Etc.

It is important that the organization think through these kinds of questions in a structured way before deciding on an approach. However, one size does not fit all. Having said that, there are some cases one can clearly put forth a recommendation:

- For startups of small to medium size, consider scrum of scrums.
- For startups that want to rapidly scale and can structure / restructure their organization without significant pushback, con-

sider LeSS.

- For existing organizations that have established processes but want to move into an Agile / Scrum development model, consider SAFe®.

Beware of experts that push only one or the other enterprise scaling model – the complexities of enterprise scaling have a lot of gray undertones representing the diversity of company situations, markets, and opportunities. There are very few crisp, black-and-white, one-size-fits-all approaches.

Last but not least, remember that underlying all enterprise scaling frameworks, you will find Agile thinking and Scrum. As such, it is possible to succeed using scrum of scrums, SAFe®, or LeSS as long as the Agile mindset remains the focus.

Chapter 5
Agile Execution Focus

Agile / Scrum frameworks and methodologies are very much focused on the here and now of delivering value. Accepting changing requirements and responding dynamically to changing needs is at the core of why teams often chose to pursue Agile / Scrum. However, all that dynamism can result in teams losing sight of what they are actually delivering, and why.

What is often ignored is that Agile / Scrum teams are highly dependent on a clear product vision, an achievable product roadmap, and a reality-based release plan.

- The **product vision** essentially defines the high level goals for the product and its alignment with the company's strategy.

This is the 30-second elevator pitch that provides the purpose and context for your product or project to anybody not familiar with your business.

- The **product roadmap** provides a holistic view of the product features that create the product vision. By its very nature, it is more granular and detailed than the vision.
- The **release plan** shows release timing for specific product functionality – what features / functionality will be available in what release. Sprint planning, team velocity, user story points, and burndown charts all contribute to making a release plan a reality.

At this level, the vision, roadmap, and release plan often are aspirational – meaning they are the planned targets to work toward. Aspirational planning targets get refined into concrete plans and deliverables by going through an Agile / Scrum process that delivers a release.

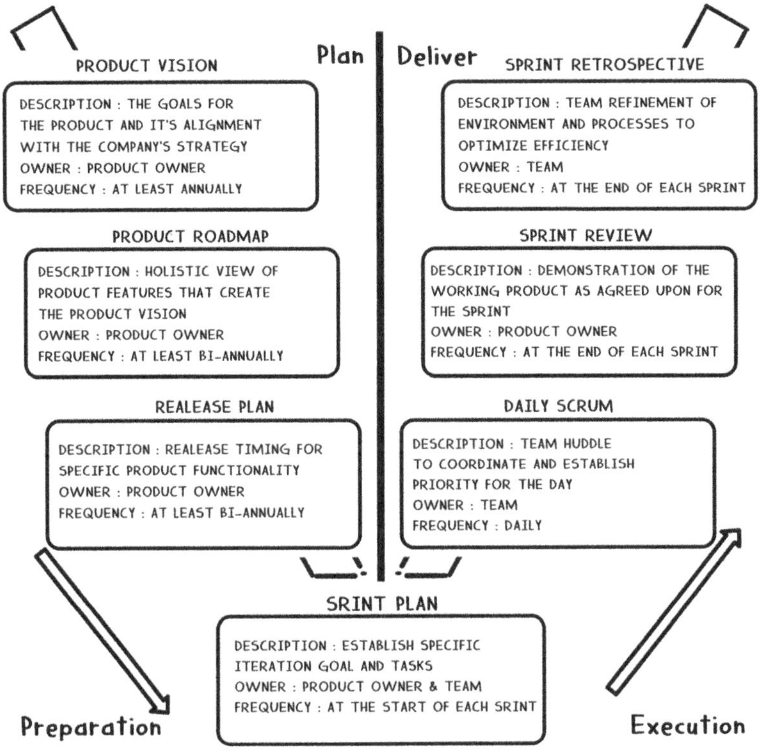

Figure 13 - Agile / Scrum Planning and Execution Phases

All three of these are the responsibility of the product owner. In large organizations, the product owner might be responsible for interfacing with departments providing them. Vision, roadmap, and release plan are **not** explicitly part of Agile / Scrum; rather, they are necessary activities outside of Agile / Scrum framework.

Once the targets are defined in the Preparation / Planning phase, it is the job of the Agile / Scrum team to deliver those targets in the Execution / Delivery phase. Aspirational targets turn into concrete deliverables through estimation, user story refinement, daily scrums, sprint reviews, and sprint retrospectives. As the team learns and refines, targets might be readjusted.

Without a clear vision, roadmap, and release plan, Agile / Scrum teams will not be successful. An Agile / Scrum team without a clear vision, roadmap, and release plan is the equivalent of a truck driver with no understanding of where he is, what he has to deliver, or how much gas is left in the tank.

This puts the product owner smack in the middle of all the delivery pressure. To demonstrate some of the pressure points, look at how Agile / Scrum fits into the picture.

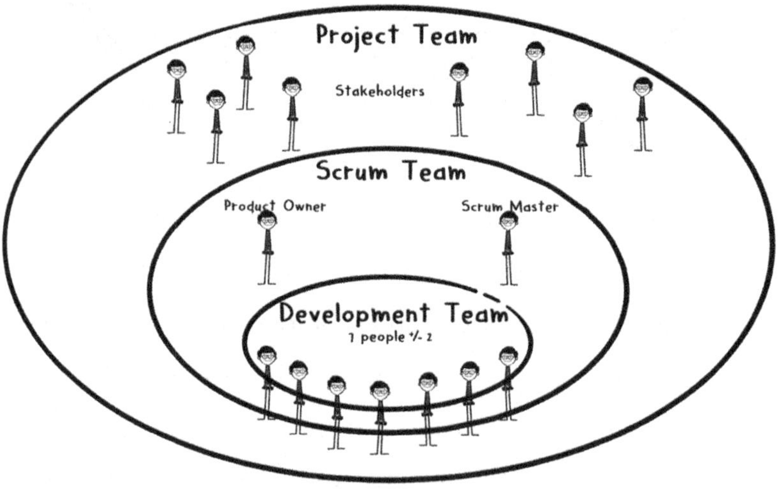

Figure 14 - Team Definitions in Scrum

We have the core development team that is supported by a scrum master and a product owner, making up the Scrum team. In some circumstances you might have an Agile coach supporting the team.

There might be one or many stakeholders that are part of the project team. For example, you might have a centralized Database or Business Intelligence team that influences your product delivery one way or another.

And then you have various influencing forces that determine one or several factors of the product. Examples include regulatory changes, competitive pressures, changing technology, budgets, etc.

The simple truth is that in order to have Agile execution focus, you must ensure that your product vision, product roadmap, and release plan are updated and communicated on a regular basis. Your Agile / Scrum team members as well as your key stakeholders should really understand what the product is all about – if not, you are in trouble.

You need to know what you are building and when you need it to make the business successful. Only if you can express those two factors will you be able to determine if you are successful.

So how to do you know that you have an Agile execution focus? Here are some telling signs:

- Everybody, including the Scrum team and the stakeholders, knows the product vision. Organizations that do this well create a consistency of messaging to the point where the VP of Development has the same elevator pitch as engineers working on the continuous integration setup. There is no doubt, at any level, what the vision is and what the product is about
- The Scrum team understands the business. *This cannot be*

stressed enough. The purpose of good user stories is to document what needs to be built, have a conversation with the product owner, and confirm the understanding. Execution-focused teams understand the business to the point where technical team members argue the validity of a business requirement with the product owner based on their understanding of the business.

- Competitor products are available to the Scrum team. Whenever teams perform well against a vision and roadmap, one thing stands out: They know their competition. You can hear discussions like "Wow, look at what they are doing!", "Wonder how they did that?", "That wouldn't work for us because feature XYZ covers that already...." Knowing your competition and how your competition implements features leads to a certain "team one-upmanship" that benefits the product.
- The team has date sensitivity. Top performing teams seem to be super sensitive to delivery dates. Especially when competing in the open market, oftentimes the team knows when their competitors delivered last, or know from trade press / trade shows what the anticipated release date or release cadence of the competitor product is. Sometimes this is done informally; sometimes the product owner or product management come back from the field to share what they found out.
- If there is seasonality that drives the business (like Thanksgiving or Christmas retail seasons, Super Bowl specials, Valentine's Day, tax season for tax software, etc.), the team under-

stands that, anticipates the impacts, and drives tradeoff decisions based on that understanding.

- The team understands "good enough." Execution-focused teams know when a feature needs to be polished and slick and bulletproof versus when something can be a little bit more clunky or "good enough." They make that tradeoff because they understand what features are critical for the business because such features might be used a million times versus what is used one time only.
- Execution-focused teams are notoriously delivery-focused – they get stuff out the door – and somewhat rebellious towards "process." If given a choice between delivering something to market on time or following a process, the process is often short-changed.
- Although the product owner is the business expert on the team, over time others pick up this business acumen. Together, they come up with better solutions – oftentimes as a result of vigorous discussions about the benefits or drawbacks of specific features. These healthy discussions are a sign that the team is working well together.
- Somebody makes the final call. Execution-focused teams have all of the above, teamwork, curiosity, business smarts, date sensitivity, open discussions, etc. But ultimately, teams that deliver also have a strong product owner who makes the call. What does not work are committee decisions. Feedback is solicited, mulled about, and discussed, but ultimately somebody

has to make the call. In Agile / Scrum, that is the product owner

Agile Execution Focus is more than just following an Agile / Scrum process. Product vision, product roadmap, and a release plan are critical to make an Agile / Scrum team execution focused. Without these critical preparation and planning activities, as well as the effective and repeated communication of them, Agile / Scrum teams often cannot live up to their full potential and truly deliver great business value within the time frame desired.

For some additional reading about execution focus in general, check out this classic: ***Execution: The Discipline of Getting Things Done***[13], by Larry Bossidy and Ram Charan.

[13] https://www.amazon.com/Execution-Discipline-Getting-Things-Done/dp/1847940684/

CHAPTER 6
SCRUM VS. KANBAN

Neither Scrum nor Kanban are new. It is worth explicitly pointing that out because many folks say something along the lines that Scrum and / or Kanban are "new development approaches". That is not true at all.

Both go back to the 1940s and 1950s as both Scrum and Kanban have roots in the ***Deming Cycle***[14], aka Plan-Do-Check-Act (PDCA) Cycle. In many ways, PDCA is the source for many modern, iterative, process improvement, and software development methodologies[15].

---

14 http://en.wikipedia.org/wiki/PDCA

15 http://en.wikipedia.org/wiki/Software_development_process

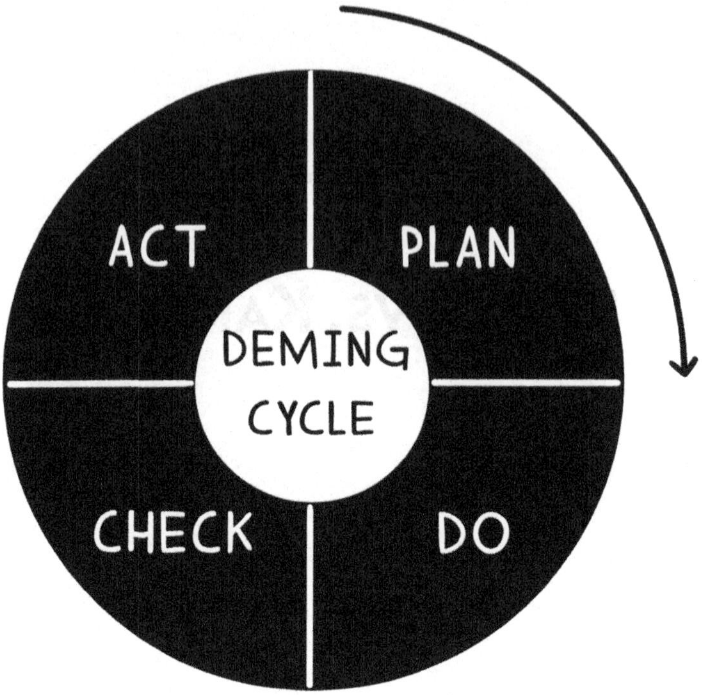

Figure 15 - Deming Cycle aka Shewhart Cycle

Scrum is an Agile framework for completing complex projects and has roots in the agile movement of the 1990s, and the resulting ***Agile Manifesto***[16]. Scrum originally was formalized for software development projects (as many of the Agile Manifesto signatories were software guys worried about making programmers more productive). Having said that, Scrum also has proven to work well for any kind of complex project work and today is used in such diverse industries such as biotech and electronics manufacturing.

Kanban, on the other hand, is a scheduling system for lean

---

[16] http://agilemanifesto.org/

manufacturing and just-in-time manufacturing. Taiichi Ohno, an industrial engineer at Toyota, developed Kanban to improve manufacturing efficiency. Kanban became an effective tool to support running a production system as a whole, and an excellent way to institute continuous improvement efforts. Kanban became part of the larger Toyota Production System[17].

The reader will find many, many opinions about what Agile, Scrum, and Kanban actually mean, and what their differences are. As with all heated methodology discussions, people pick sides and defend their viewpoint as if their life depended on it.

Just to make the point, here is a listing of Agile software development frameworks / methodologies one can find by googling "Agile Frameworks List":

- Adaptive Software Development (ASD)
- Agile Modeling
- Agile Unified Process (AUP)
- Crystal Clear Methods
- Disciplined Agile Delivery
- Dynamic Systems Development Method (DSDM)
- Essential Unified Process (EssUP)
- Extreme Programming (XP)
- Feature-Driven Development (FDD)
- Lean Software Development

---

[17] http://en.wikipedia.org/wiki/Toyota_Production_System

- Open Unified Process (OpenUP)
- Kanban
- Scaled Agile Framework (SAFe®)
- Scrum
- Scrumban

With less than 30 minutes of research anybody could double the list. Agile methodologies / frameworks seem to multiply on a regular basis.

This chapter will focus on the differences between Scrum and Kanban, as those two terms seem most prone to confusion and conflicting viewpoints. A good start is describing the big differences and then diving into some details, fleshing out the nuances.

HIGH LEVEL DIFFERENCES BETWEEN SCRUM AND KANBAN

	Scrum	Kanban
Delivery Cadence	Regular **fixed length sprints** (e.g., 2 weeks)	Continuous flow
Release Methodology	At the **end of each sprint** if approved by the product owner	**Continuous** delivery or at the team's discretion
Roles	Product Owner Scrum Master Development Team	**No** predefined roles
Key Metrics	Velocity	Cycle Time
Change Philosophy	Teams should strive to **not make changes** to the sprint forecast during the sprint. Doing so compromises learnings around estimation.	Change can happen at **any** time

Table 1 - Differences between Scrum and Kanban

In short, Scrum is focused on delivering releases at predetermined intervals (sprints) vs. Kanban's focus on continuous delivery. This key difference explains why many software companies develop their products or services using a Scrum model, but support their products or services by following a Kanban model.

As mentioned in the previous chapter, Scrum is a framework used to organize work into small, manageable pieces that can be completed by a cross-functional team within a prescribed time period (sprints, generally 2-4 weeks long).

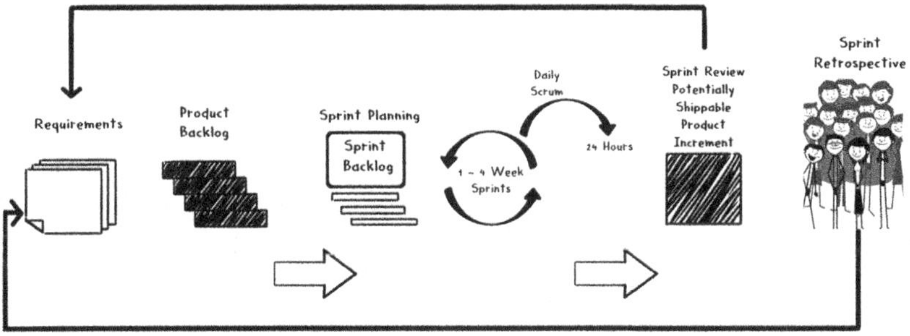

Figure 16 - Scrum Overview

To plan, organize, administer, and optimize this process, Scrum relies on at least three defined roles: the product owner (responsible for initial planning, prioritizing, and communication with stakeholders), the scrum master (responsible for overseeing the process during each sprint), and the development team members (responsible for carrying out the purpose of each sprint). In some circumstances an Agile coach assists with the transition to from an existing process to Agile / Scrum.

Scrum uses **scrum task boards** – a visual representation of the work flow, broken down into manageable pieces of work called **user stories**, with each user story moving along the board from the **sprint backlog** (the to-do list), into work-in-progress (WIP), and on to completion.

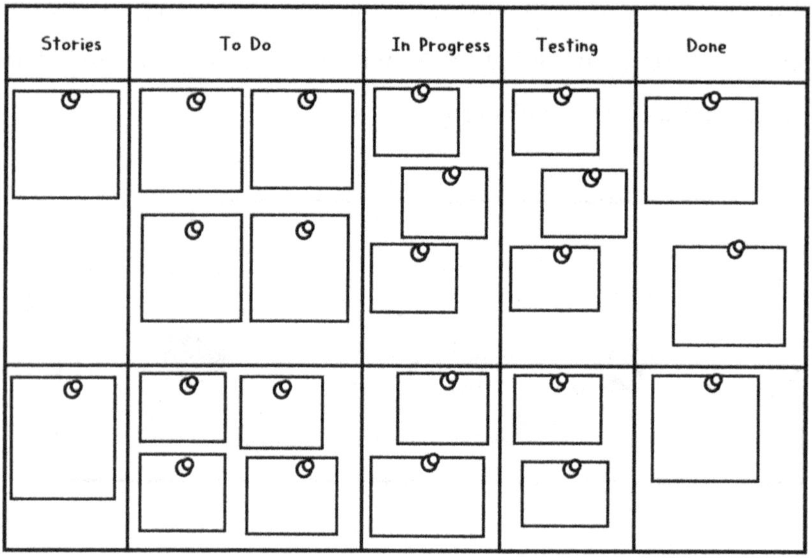

Figure 17 - Scrum Task Board

Finally, Scrum measures progress by velocity and tracks progress in **burndown charts**.

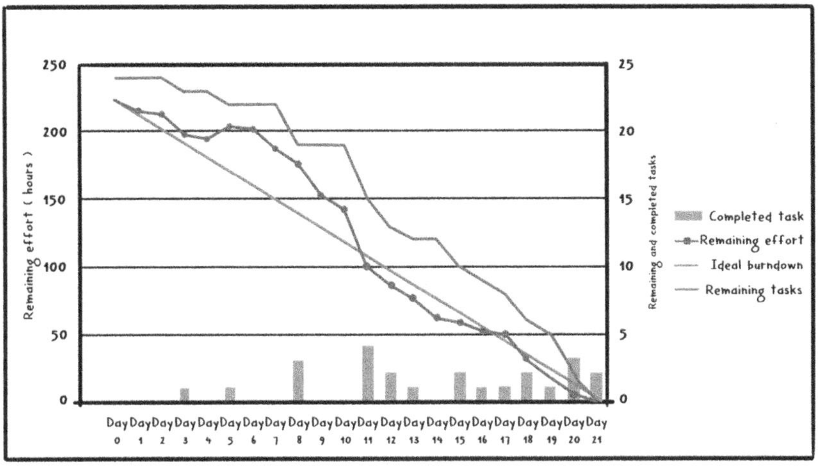

Figure 18 - Scrum Burndown Chart

Kanban, on the other hand, is a tool that is based on what is commonly called the "pull system". The question then becomes "What is a pull system"?

Figure 19 - Push vs. Pull Model

It's important to remember that Kanban and the original work done by Toyota was largely focused on optimizing work on a physical production assembly line (as opposed to abstract software development).

Because Japan was extremely resource-limited after WWII, the production philosophies of American and Japanese production lines were opposite of each other:

- **American production lines** tried to produce as much as possible as quickly as possible, counting on the fact that ultimately demand would compensate for any inefficiencies (remember, America had come out of the Great Recession and the post-WWII boom offered unlimited growth opportunities based on pent-up consumer demand).
- **Japanese production lines**, on the other hand, tried to optimize resources by carefully pulling each item to be built through the system – a consequence of the extreme resource scarcity experienced post-WWII and the general lack of demand – large lots could not be guaranteed to be bought up by pent-up consumer demand like in the US.

The source inspiration for Kanban and the "pull system" was the supermarket. Store clerks restocked a grocery item by their store's inventory, not their vendor's supply. Only when an item was near sell-out did the clerks order more. The grocers' "just-in-time" (JIT) delivery process sparked Toyota engineers to rethink their methods and pioneer a new approach — a Kanban system — that would match inventory with demand and achieve higher levels of quality and throughput.

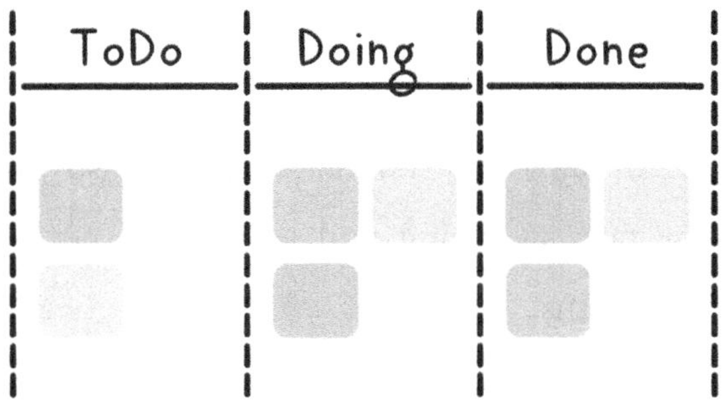

Figure 20 - Kanban Board

Kanban is Japanese for "visual signal," "visual limiter," or "card." Toyota line-workers used a Kanban (i.e., an actual card) to signal steps in their manufacturing process. The system's highly visual nature allowed teams to communicate more easily on what work needed to be done and when. It also standardized cues and refined processes, which helped to reduce waste and maximize value.

Kanban helps harness the power of visual information by using sticky notes on a whiteboard to create a "picture" of your work. Seeing how your work flows within your team's process lets you not only communicate status but also give and receive context for the work.

There are Four Core Kanban Principles:

1. **Visualize Work** – Visualizing your work model lets you ob-

serve the flow of work moving through your Kanban system. Better visibility instantly leads to increased communication and collaboration.

2. **Limit Work-in-Progress (WIP)** – By limiting how much unfinished work is in progress (meaning somewhere in the production pipeline), you can reduce the time it takes an item to travel through the system. You can also avoid problems caused by task switching and reduce the need to constantly reprioritize items.
3. **Focus on Flow** – By using work-in-progress (WIP) limits, you can optimize your system to improve the smooth flow of work; you can collect metrics to analyze flow and improve throughput.
4. **Continuous Improvement** – Once your Kanban system is in place, it becomes the cornerstone for a culture of continuous improvement. Teams measure their effectiveness by tracking flow, quality, throughput, lead times, and more. Experiments and analysis can change the system to improve the team's effectiveness.

Differences between Scrum Task Boards and Kanban Boards

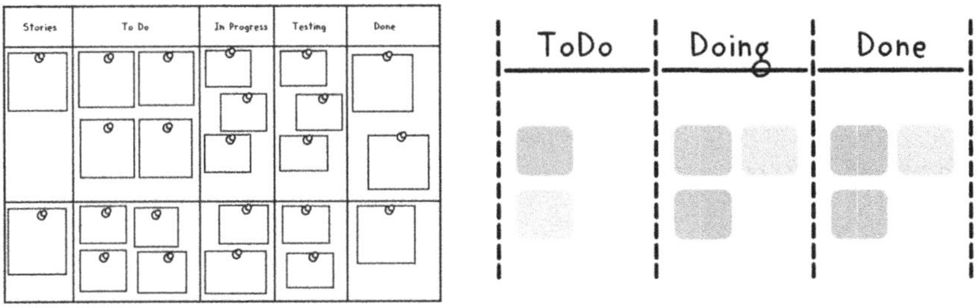

Figure 21 - Scrum Task Board vs. Kanban Board

Scrum task boards and **Kanban boards** look suspiciously similar, which is primarily the reason so many people are confused about the differences. Here are some of the differences:

Scrum Task Board	Kanban Board
Scrum task board is a board tracking work in sprints. A sprint is a short, consistent, and repetitive period of time (usually more than a day, but less than 4 weeks).	Kanban board is a board tracking the process flow while maintaining the number of work-in-progress activities. The amount of work-in-progress is small enough to avoid unworthy tasks but big enough to reduce idle personnel.
Scrum is like a school test: you have to complete a set of tasks in a certain period of time. And no other activities are allowed.	Kanban is like a basketball match, where a completed task equals one point and a team is trying to minimize the time between shots.
Scrum limits work-in-progress per iteration. The team has to commit to the number of tasks that they are ready to accomplish during the sprint. The Scrum team ultimately decides what is right for the sprint.	Kanban limits work-in-progress per workflow state in order to avoid wait / hold times.
Scrum task board is always owned by one cross-functional Scrum team responsible for the successful completion of all the tasks during this sprint.	Kanban board is not owned by a specific team, since it is mostly devoted to a workflow. In reality, some Kanban boards will outlive the teams it is serving. Workflow is the focus.

Scrum Task Board	Kanban Board
On Scrum task board, the whole team is devoted to each task. The team does not succeed until every task on the board is done according to the "definition of done." One for all, and all for one.	On Kanban board, a team is not originally responsible for all the tasks. Each person is responsible for his/her step on the task flow. However, Kanban has a culture of slack resources to help resolve bottlenecks, when needed. Slack resources are any non-bottleneck resource. So if a person has completed his task while there's something complicated in testing, he can choose what to do: help the tester to complete the task or take another activity from the queue.
The scope of work in a sprint is fixed once agreed upon and once the sprint is started. "In-sprint" changes are to be avoided. The number of items is set during planning session, before the sprint starts.	There are no timeframes for updating a Kanban board (the team estimates the flow using historical data), because it limits the work in progress activities. As soon as any task moves from the In Progress column to the Done section, and the capacity is released, the new item goes under development.
Unexpected, urgent, work items are rarely added to the sprint. One main idea of Scrum is to allow team members to be optimally productive by minimizing interruptions and getting into the "flow" of things.	Some teams add an Urgency section on Kanban board, represented as a swim lane where it's supposed to get a maximum speed. It can be an unpredicted urgent task from the backlog or a bottleneck task from the board.
Scrum considers user stories as the best practice to create a product backlog	Kanban also uses a backlog practice; the format is open to the specific application space.
If the Scrum team commits to items, then they estimate them as achievable within the sprint timeframe. If a user story is too big for the sprint it needs to be split into smaller user stories until it fits the sprint time frame.	There is no specific rule regarding the volume of the task in Kanban.
In Scrum, prioritization is a must. Most important items are prioritized to the top of the stack, least important items to the bottom of the stack. This sorting and	Kanban does not use a prioritization or estimation scheme, but considers project planning using probabilistic forecasting.

Scrum Task Board	Kanban Board
grooming and refining of the product backlog, setting priorities and resource estimation, guarantees that items with the most business value are worked on first. During prioritization it is important to predict what will be important for the next (not the current) sprint.	
Scrum is using a velocity as the primary metric, accompanied with a range of charts and reports: sprint burndown chart, release burndown, velocity chart, etc.	In Kanban, particular charts are not prescribed. Various reports can be used as needed, for example a lead time report.
At the end of each sprint all the items should be in the final Done section; otherwise the sprint is considered as unsuccessful. After the quick retrospective, all the stickers are removed to clear the board for the next sprint.	A Kanban board is a persistent tool with no timeframes, so it is not necessary to reset it and start over. The flow continues with the project life cycle, and the new items are added as soon as the need arises.

Table 2 - Detailed Comparison of Scrum Task Board and Kanban Board

Kanban is a tool used to organize work for the sake of efficiency.

Like the Scrum framework, Kanban encourages work to be broken down into manageable chunks (called user stories in Scrum). Both Kanban and Scrum use boards to visualize their workflow, but Kanban boards "flow" endlessly without time limitation, whereas Scrum task boards are time-boxed to the sprint duration. Scrum limits the amount of time allowed to accomplish a particular amount of work by time-boxing sprint durations; Kanban focuses on limiting the amount of work that flows through any one state.

Scrum, although used in many different industries, originated in software development. Consequently, the focus was to put an efficient framework around somewhat intangible software development processes and address the bane of software development: endless schedule slips. Time-boxing delivery cycles is at the core of Scrum.

Kanban, on the other hand, originated in the physical world of production assembly lines. Therefore, Kanban focuses on determining how an assembly line can be optimized, and how pieces can optimally be put through the line, guaranteeing optimal resource use (materials, people, and machinery). Kanban later was adapted to serve other problem spaces, including software and service delivery.

Both Scrum and Kanban allow for large and complex tasks to be broken down and completed efficiently. Both place a high value on continual improvement, optimization of the work and the process. Both share the similar focus on making work visible in order to keep team members in the loop.

Scrum is mostly focused on releasing product in predetermined time frames (sprints), while Kanban is focused on continually delivering throughput based on a pull principle.

CHAPTER 7
AGILE RELEASE PLANNING

Release planning using Agile / Scrum is easy!

A statement like that usually gets interesting reactions from an audience, from disconcerted, basically expressing the thought that obviously the person uttering such thoughts has no clue what he is talking about, to panicking, exposing the inner fear that there might be some secret knowledge they cannot even comprehend.

Agile / Scrum release planning is easy – if you just follow a couple of simple best practices.

This chapter summarizes the key points of **Agile release planning** in just a few pages. The goal is to give you a primer you can fall back on during your release estimation session.

To get the key points across, many illustrations are used.

The Inevitable Question

Imagine that you are working in a software company that just released its first product, *BubbleSoft Platform 1.0.* Sooner or later, you will have a conversation like this:

"Our product, BubbleSoft Platform, just released version 1.0. Great job guys! We are selling like hotcakes, the press loves us, and the demand is far greater than anticipated! Off the charts! That's the good news.

Bad news is that we need to know by next Tuesday when we can release version 2.0, and we need to tell the market what the target release date for version 3.0 will be as well.

The reason for this is that we are trying to put together marketing and tradeshow budgets and need to know what we can show at next years' ABC tradeshow.

Oh, and we are looking for next round funding and the venture capital guys ask us for those dates as well, because they want to understand our 'burn rate,' whatever that is... ***Can anybody give me the dates?***"

There it is, the dreaded question! In many organizations, this question causes a lot of consternation, followed by frantic estimation exercises trying to discern a somewhat realistic date.

Usually this is based on wild guesswork because requirements and specifications have not been written yet. This in turn leaves the engineers trying to guestimate as best they can based on their understanding of what the requirements might be.

Once the engineers come up with a timeline, executives might push back because they fear that timelines are unacceptable to the market, their venture investors, or their budgets. At the other extreme, executives might buffer imaginary schedules because they do not trust the estimates to begin with.

The point is this kind of estimation effort frequently results in fantasy schedules because they are built on quicksand: no or insufficient requirements, unknown scope, hasty estimation effort.

In an Agile / Scrum environment on the other hand, you can answer this question relatively easily, assuming you have several key factors taken care of:

1. You have a **vision statement** that describes your product or project.
2. You have a well-maintained **product backlog,** including well-defined user stories.
3. You have **sized** the product backlog items, ideally using estimation poker and story points, but if the volume of product backlog items is too great, at a minimum you have T-shirt-sized stories that are not in the immediate next two sprints.
4. Your **product owner** is able to assign specific product backlog items to specific releases using a product release roadmap.
5. You have a clear **definition of ready** and **definition of done.**
6. You understand your team's **velocity**.

WHAT IS A VISION STATEMENT?

A **vision statement** is an elevator pitch that puts the product or project goals in the context of the marketplace and your customer needs.

Customers can be either internal (say for an IT project) or external (to be sold or made available for free to people outside of your organization).

It is strategic in nature and owned by the product owner. For a vision statement to be effective, it must be communicated throughout the development and supporting organizations in order to set expectations.

Simply put, the vision statement provides context for how the current development effort supports the overall goal.

Vision statements are usually pretty short. ***Geoffrey Moore ("Crossing the Chasm")***[18] suggested the following phrasing format:

> **For** <target customer> **who** <statement of the need> **the** <product name> **is a** <product category> **that** <product key benefit, compelling reason to buy>.
> **Unlike** <primary competitive alternative>, **our product** <final statement of primary differentiation>, **which supports our strategy to** <company strategy>.

Pretty straightforward and to the point. Please note that there are other ways to create a vision statement, such as Bill Shakelford's 'Design the Box' exercise.

---

[18] http://www.amazon.com/Crossing-Chasm-3rd-Disruptive-Mainstream/dp/0062292986/

WHAT IS A PRODUCT OR PROJECT?

A vision statement usually initiates team members to the purpose of the product or project at an elevator pitch level of granularity. A product or project is a development effort that can be broken down into packages of progress.

The first package of progress is *timing* focused. A product breaks down into different releases, which in turn break down into sprints, which in turn are made up of weeks that have several days.

According to Scrum, a sprint can be as little as 1 day or as long as 4 weeks in duration, so weeks is an optional package, or only relevant if the sprint duration exceeds 7 days.

Please note that products are usually "externally sold," whereas projects often are "internal efforts." Think "BubbleSoft Platform" sold at $1,999 per seat versus an internal IT project called "Windows 10 Upgrade / Refresh Effort."

Products by their very nature are longer term solutions that lend themselves to feature / functionality delivery over multiple releases, whereas projects can be multi-phase but often are one-shot efforts.

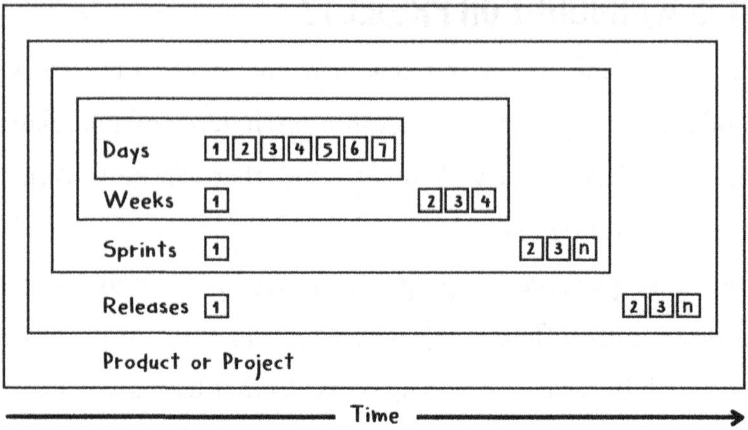

Figure 22 - Continuously Refined Packages of Progress – Timing Focused

An alternative view looks like this:

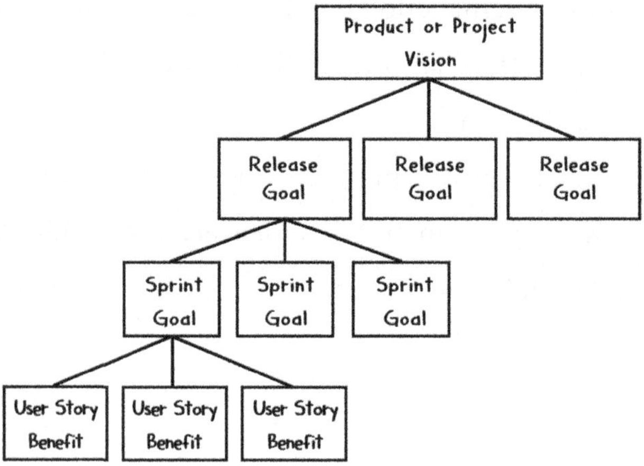

Figure 23 - Continuously Refined Packages of Progress - Goal Focused

The second package of progress is *goal* focused. The high level vision translates into release goals that are further broken down into sprint goals, which finally break down into specific user story benefits.

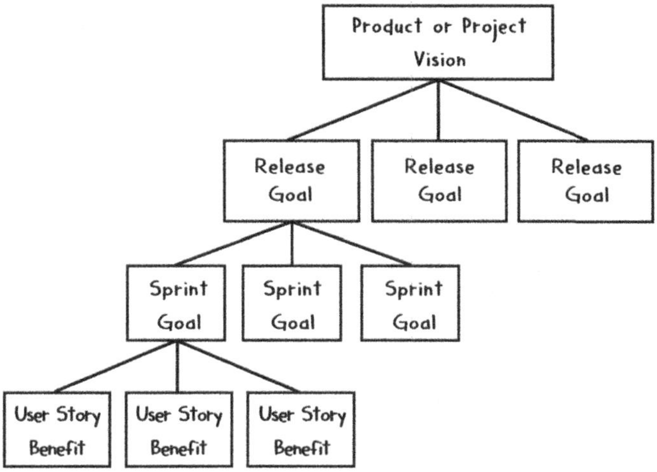

Figure 24 - Continuously Refined Packages of Progress - Scope Focused

The third package of progress is *scope* focused. The high level vision breaks down into a roadmap describing features, which are further broken down into epics, which break down into specific user stories and associated tasks.

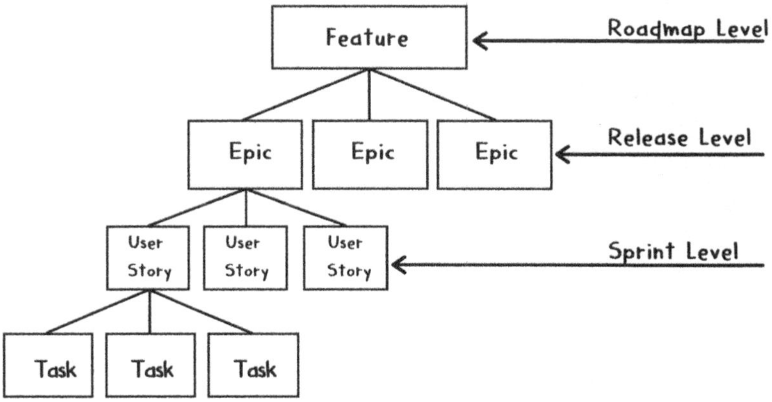

Figure 25 - Planning Horizons - Roadmap, Release, Sprint Levels

Now that we identified the vision statement, the product, and the package structure of progress (timing, goal, and scope focus), the question becomes how to manage all of this? The answer is the **product backlog.**

THE PRODUCT BACKLOG

The **product backlog** is the product's or the project's **to-do list.** All scope items, regardless of level of detail, are in the product backlog.

It is ordered, not just prioritized; product backlog items are prioritized according to business value. What is known is written down and documented, with the expectation that discovery will lead to change. The product backlog is owned by the product owner.

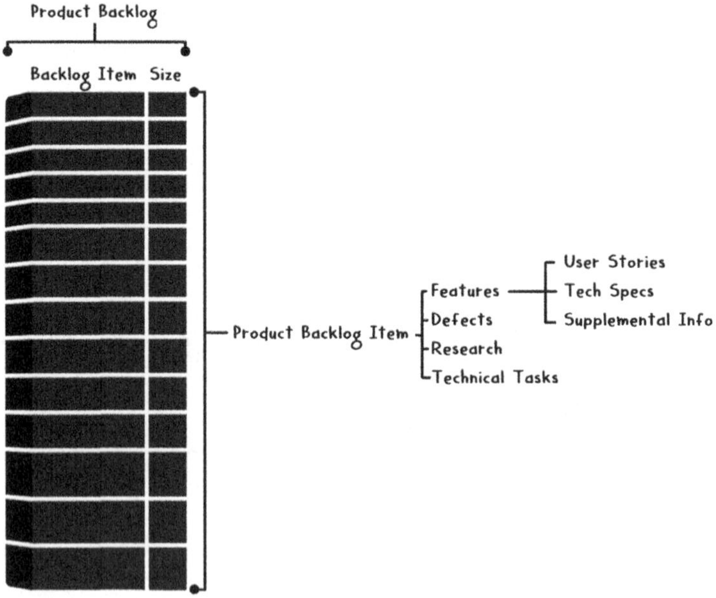

Figure 26 - The Product Backlog

The product backlog is a living document / repository and is expected to change.

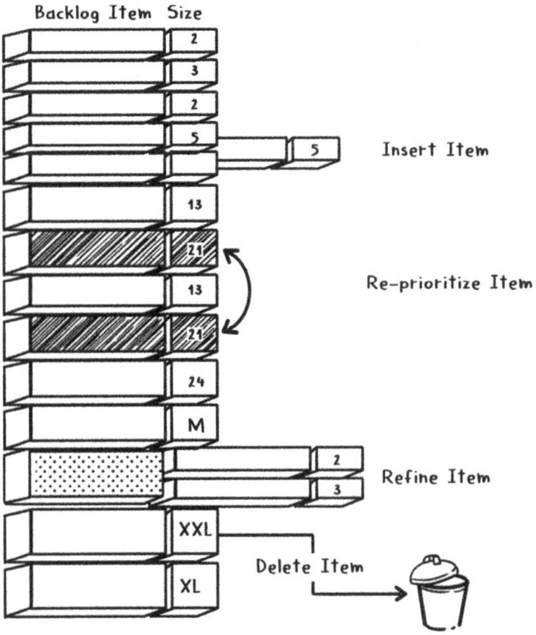

Figure 27 - Product Backlog Grooming / Refinement Process

The product owner is expected to insert, re-prioritize, refine, or delete items from the product backlog. This can happen any time until the sprint scope is defined and committed to by the development team.

Sizing and estimation of the product backlog usually occurs in sprint planning meetings or at regularly intervals during ongoing sprints. Depending on how fast the product owner adds user stories, more frequent — maybe even daily — estimation sessions might be needed.

The important point here is that product backlog items need to be estimated continuously in order to avoid the big bang estimation effort at one critical juncture of the effort. This has the added benefit of exposing the entire team to the product owner's thinking, keeping everybody in sync and with an improved understanding of requested product features and functionality.

Sprint planning meetings or more frequent estimation session need to follow established Agile estimation best practices. Product backlog items are estimated using either estimation poker or T-shirt sizing. To see how to effectively use these estimation methods, check the corresponding chapter.

In order to have a somewhat comparable scheme between estimation poker and T-shirt sizing, it's a good idea to use the following quantification scheme:

- XS = 1 story point
- Small = 2 story points
- Medium = 3 story points
- Large = 5 story points
- XL = 8 story points
- XXL = 13 story points
- EPIC = Product owner needs to break down the epic into more refined user stories. (For example, the epic might state something like "internationalization support." To do a proper estimate, this needs to be broken down into more detailed user

stories, such as "provide support for British English" or "provide support for Spanish.")

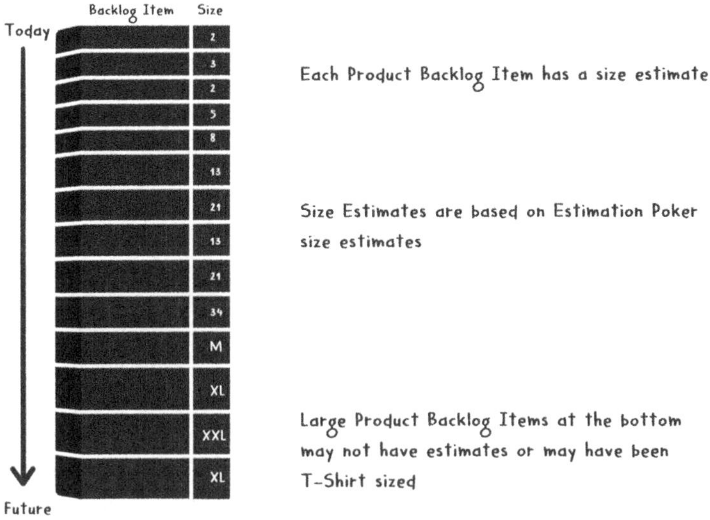

Figure 28 - Product Backlog Sizing

The product backlog is the source of the **sprint backlog**. Once a sprint goal / scope has been committed to by the development team, sprint backlog items should not be changed.

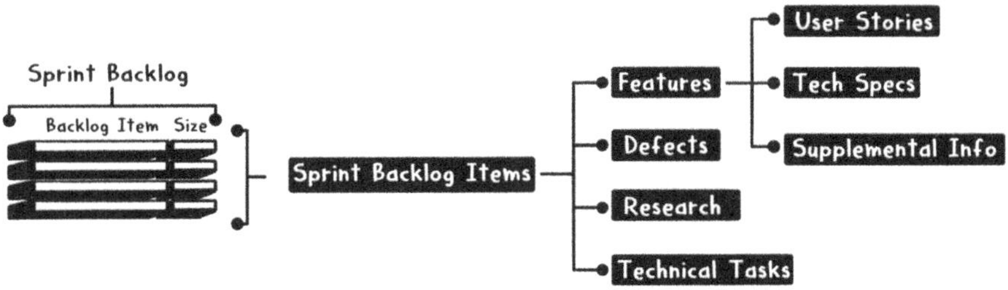

Figure 29 - The Sprint Backlog

AGILE RELEASE PLANNING IS BASED ON A PRODUCT ROADMAP

A **product roadmap** represents a holistic, yet digestible, view of features that enable the product vision.

It enables affinities to be established and gaps to be identified, and in general shows release timing for product features. Highest priority features are released first. It is produced by the product owner and prepared at least bi-annually.

Date	1st Quarter	2nd Quarter	3rd Quarter
Name	BubbleSoft Platform 1.0	BubbleSoft Platform 2.0	BubbleSoft Platform 3.0
Goal	Free App, Limited Purchases	In-App Purchases	Retention
Features	Basic Functionality Multi Language Support PayPal Integration	Chinese Language Support Visa / Master / Amex Integration Purchase Bubbles	Defect Fixes / Stability Enhancements Gamification iOS 11 Support
Metrics	Downloads in top 10	Activations Downloads	Downloads

Figure 30 - Sample Product Roadmap

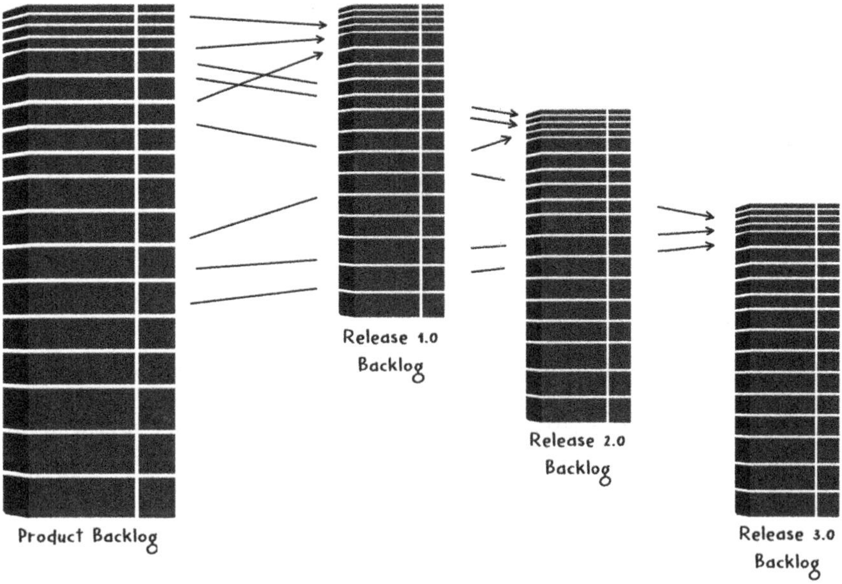

Figure 31 - Product Backlog feeding various Release Backlogs

AGILE ESTIMATION IS BASED ON EMPIRICAL DATA – NOT GUESSWORK!

What is **velocity**? Velocity is the rate at which the development team can reliably deliver story points (usually packaged into time-boxed sprints and resulting in a working product increment).

Definition	1. Rapidity of motion or operation; swiftness; speed: a high wind velocity. 2. The time rate of change of position of a body in a specified direction. 3. The rate of speed with which something happens; rapidity of action or reaction. 4. Points delivered by a team per sprint.
How to Calculate	Rolling average of sprints Maximum Minimum Lifetime average
How to Use	To determine sprint scope.

	To calculate approximate cost of a release. To track release progress.
Remember	Velocity of two teams is not comparable. Velocity changes when the team composition changes. Velocity increases with team's tenure. Velocity does not equal productivity. Defects and rejected stories count against velocity!

Table 3 - Velocity Definition

Using estimation poker, estimation variability decreases usually over the course of the first 4 to 6 sprints. It is rare to see stable teams that do not achieve fairly accurate estimates after working together for 5 or 6 sprints.

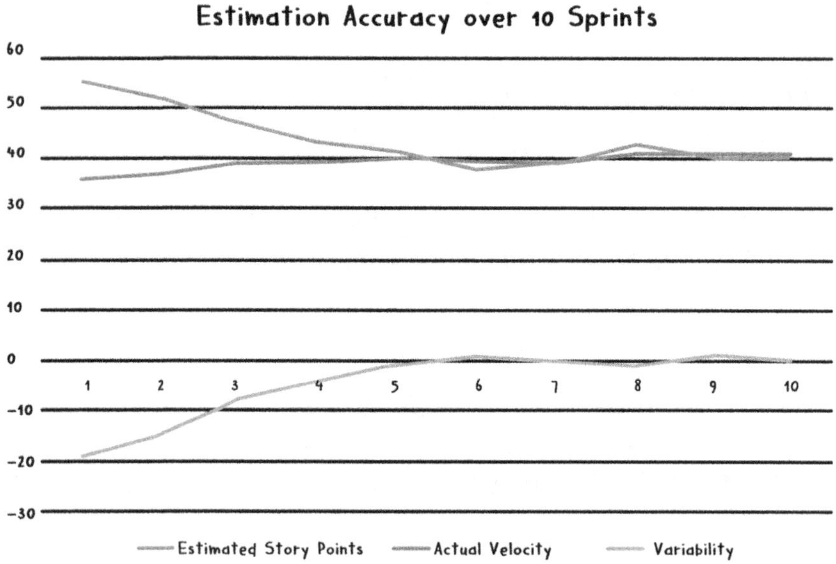

Figure 32 - Improving Estimation Accuracy over Time

Pleases note that "stable teams" is key here – teams need to be stable, meaning team members need to know each other, have worked with each other, and successfully formed a team – following the standard "forming–storming–norming–performing" model of group development, first proposed by Bruce Tuckman in 1965.

Once a development team has gone through the "forming–storming–norming–performing" process, team velocity is established. Estimation accuracy and predictable velocity allow for longer term forecasting.

AGILE RELEASE PLANNING SAMPLE

For example purposes, if your product backlog contains 450 user stories representing 2,400 story points and your development team is delivering at a steady velocity of 60 story points for every 2-week sprint, you are able to predict that the rest of the project will take another 40 sprints, which equals 80 weeks, or roughly 1 ½ years.

- 2,400 / 60 = 40 sprints
- 40 sprints × 2-week sprint duration = 80 weeks
- 80 weeks / 52 weeks in a year = 1.54 years

Accordingly, calculating your burn rate becomes a trivial exercise. Simply multiply your labor cost and overhead by the duration. For example, if you know that your 7-member development team costs you $1,529,500 per year ($805,000 for salaries, $724,500 for overhead and fringe benefits, etc.) then you can easily figure out the cost per sprint ($1,529,500 / 26 = $58,827).

You cannot get any better at forecasting and predictability.

Estimation poker is an empirical, self-correcting, approach to short-term software estimation where user stories are fully defined.

T-shirt sizing (affinity estimation) is used for higher level estimates needed for budgeting, staffing, and other longer term planning activities where not all user stories are defined in detail and a quick assessment of upcoming work is needed.

Returning to our original question "***...we need to know by next Tuesday when we can release version 2.0, and we need to tell the market what the target release date for version 3.0 will be as well.***" Let us quickly figure out when the 2.0 and 3.0 releases will be able to ship.

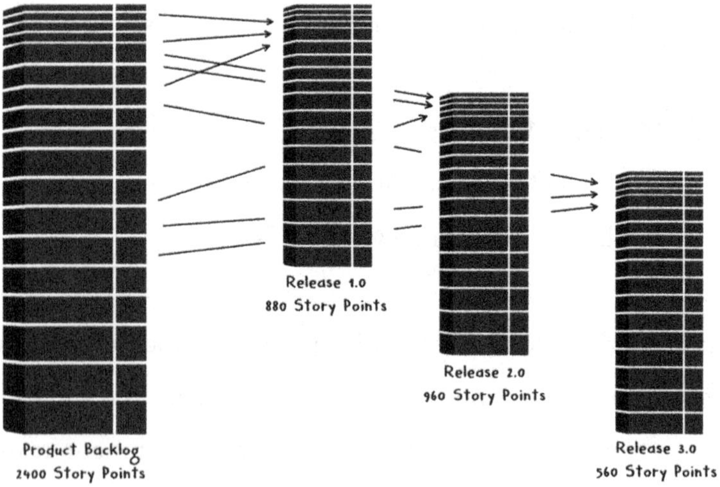

Figure 33 - Product Backlog feeding various Release Backlogs with Story Points

We released the 1.0 version of our product, BubbleSoft Platform, which consisted of 880 story points. That is done. The release provided us with a stable development team that delivered at a predictable velocity of 60 story points per sprint (2-week sprint duration).

Based on our existing velocity of 60 story points, we now can calculate how long it will take us to deliver the remainder of the product backlog, which the product owner assigned to version 2.0 (960 story points) and version 3.0 (560 story points).

Release	Story Points	Velocity	Sprints	Weeks	Months	Cost per Sprint	Cost per Release
1.0	880	60	14.7	29.3	6.8	$58,827	$862,796
2.0	960	60	16.0	32.0	7.4	$58,827	$941,232
3.0	560	60	9.3	18.7	4.3	$58,827	$549,052
	2,400		40.0	80.0	18.6		$2,353,080

Table 4 - Release Plan at Steady 60 Story Point Velocity

Assuming that the development team will improve and gain efficiencies, you might expect to improve on velocity over time, which could look like this:

Release	Story Points	Velocity	Sprints	Weeks	Months	Cost per Sprint	Cost per Release
1.0	880	60	14.7	29.3	6.8	$58,827	$862,796
2.0	960	65	14.8	29.5	6.9	$58,827	$868,830
3.0	560	70	8.0	16.0	3.7	$58,827	$470,616
	2,400		37.4	74.9	17.4		$2,202,242
Savings							$150,838

Table 5 - Release Plan Assuming Gradual Velocity Increase

The increased velocity shortened the delivery time frame from a total of 80 weeks down to about 75 weeks, resulting in a savings of about $150,838.

Release planning, using Agile / Scrum, is easy. And, you cannot get any better at forecasting and predictability.

COMMON ISSUES AND CHALLENGES

The section addresses some common issues and challenges that make Agile release estimation more difficult than it has to be.

No Vision Statement

Product or project efforts frequently do not have a **vision statement**. The purpose of a vision statement is to communicate the high level goal of a product or project. Without it, team members frequently have no context for why they have to do certain work.

This is the equivalent of a truck driver knowing how to turn right or left, but having no idea what the final destination is. Vision statements are important in order to rally everybody around a common cause and to provide that "true north" compass.

No Product Release Roadmap

Roadmaps are important because there is more to software development than simply a stream of software releases. **Product roadmaps** take into account market specifics, from seasonality of the product (watch out for the proverbial Christmas trees in June!), over competitive pressures (oh, our competitors just released stuff we were planning to do next year!), to external factors (like the yearly October-ish release of iOS).

Roadmaps are essential planning tools to lay out release timing for product features. Not having one means you are flying blind.

Product Roadmap (feature desires) and Velocity Calculations (delivery capabilities) Do Not Match

Sometimes product roadmaps perfectly express the desired release timing, but your development team's velocity just cannot get you there.

From this problem, you might see many dysfunctions develop. For example, one executive team simply might not admit that the desired release date is unrealistic, overriding the development team's estimates and forcing the team to bulk up on additional developers (the old fallacy, just double the staff to cut the needed time in half!).

The result can be a bloated team that is less productive, more expensive, and completely burned out after a death march.

Remember that the good thing is that the velocity is based on empirical data. However, you cannot bend the reality curve to your liking by simply doubling or quadrupling staff.

User Stories are Inadequate

The biggest problem when it comes to user stories, bar none, is their size. Ninety percent (90%) of all problems encountered are based on user stories being too big. (See the chapter "Quick & Dirty Guide to Writing Good User Stories.")

Writing good user stories is essential in order to do a good job during the estimation poker or T-shirt sizing exercises. Inadequate or poorly written user stories make the downstream estimation process less reliable, and consequently put the entire development effort at risk.

Estimation Is Not Done on a Regular Basis

If the team does not perform estimation activities on a regular basis, then the risk becomes an excessively big backlog. You do not know if you are dealing with a molehill or Mount Everest. Therefore, you cannot plan your release schedule.

In the ideal case you would have time to estimate all user stories using estimation poker. If you cannot do that, either because the user stories do not have sufficient detail or your backlog is too big, then use T-shirt sizing.

It is not uncommon for a productive product owner to crank out hundreds and even thousands of user stories. When you have 2,986 user stories at varying levels of detail, some fully fleshed out, and some only with a subject line, then T-shirt sizing is the only viable option. After all, you do not want to put your development effort on hold for 6 weeks simply to generate estimates with estimation poker.

The drawback is that T-shirt sizing is not as precise as estimation poker – so you have to think about how to deal with estimation variance.

The Development Team Is Unstable

There are many reason why a development team might be unstable. New technologies might lure team members into new jobs. Better benefits might entice people to leave. Use of contractors might shorten the tenure on the team. Personality conflicts might cause team members to leave. People get sick. People retire. The business might decide to outsource development activities. Your outsourcing partner experiences excessive turn over which you have no influence over. Etc., etc.

Having a stable development team is by far the biggest productivity enhancer and guarantees that velocity becomes predictable over time.

If your development team is unstable, most likely you will see that your velocity becomes unpredictable.

Velocity Is Unpredictable

Stable development teams make for very predictable velocity. Unstable development teams make for unpredictable velocity.

Only estimating using T-shirt sizing introduces variance that ultimately gets expressed in unpredictable velocity. Detail level estimation using estimation poker increases precision.

User stories that are missing, too big, or poorly written make any kind of estimation difficult, and often result in unpredictable velocity.

The question is, how do you deal with unpredictable velocity in your release planning efforts?

The following table shows an example of varying velocity:

Sprint	1	2	3	4	5	6	7	8	9	10	11	12	13	14	15	16	17	18
Planned Velocity	6 0	6 0	6 0	6 0	6 0	6 0	6 0	6 0	6 0	6 0	6 0	6 0	6 0	60	40	0	0	0
Actual Velocity	58	50	63	55	58	67	41	43	57	39	47	61	59	62	39	41	20	20
Running Average Velocity	58	54	57	57	57	59	56	54	55	53	53	53	54	54	53	53	51	49
Story Point Deficit	2	12	9	14	16	9	28	45	48	69	82	81	82	80	81	40	20	0

Table 6 - Sample Varying Velocity

This table uses the same assumptions as discussed before – 60 story points velocity per sprint, and a total sprint backlog of 880 story points for the release, BubbleSoft Platform 1.0.

Based on these assumptions we should have wrapped the release in sprint #15.

But, as you can see, the actual sprint velocity was different from the planned velocity, resulting in a varying story point deficit over the course of the sprints, and resulting in a deficit of 81 story points at the end of sprint #15.

It then takes another 3 sprints to work off the deficit. Assuming our 2-week sprint duration, that's a delay of 6 calendar weeks.

Here is another view of the data in the table:

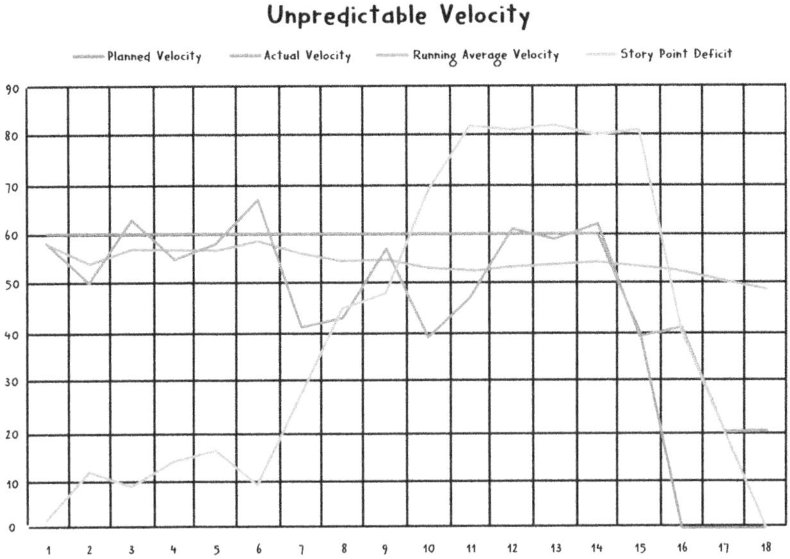

Figure 34 – Graph Showing Unpredictable Velocity

In terms of "slip," we basically missed the mark by 81 story points, which is about 9% of the overall development effort of 880 story points.

But, for whatever reason the team productivity really took it on the nose in sprints #7, #8, #10, and #11. This kind of thing can happen because of team members leaving the team and not getting replaced in time or because of illness or other reasons.

Also, it looks like the team really lost steam in sprints #17 and #18, only delivering 20 story points toward the end. This resulted in a calendar time slip of 20% (6-week slip on a 30-week schedule, which originally assume 15 sprints at 2-week sprint duration).

Again, this kind of thing can happen when engineers leave the team before things are done, or are moved on to other development efforts before the product actually ships – leaving a smaller team to mop up remaining functionality.

Regardless of the specific reasons, at the end of sprint #10 (running a 69-point deficit) it is necessary to sit down with the product owner and discuss options:

1. Deprioritize certain user stories to get the "ask" in line with the delivery capability – basically lowering the total amount of story points in the release from 880 to, say, 800.
2. Planning on additional sprints to make up for the lower than expected velocity.
3. A combination of both.

Release planning using Agile / Scrum is easy! But, it assumes you follow best practices for:

1. Laying out a **product vision,**
2. Maintaining good user stories in a **product backlog,**
3. Assigning them to a **product roadmap,** and
4. Actively **managing the release** as you progress through sprints toward your release goal.

Predictability does not come for free.

CHAPTER 8
MANAGING THE PRODUCT AND SPRINT BACKLOGS

One of the most frequent challenges for Agile / Scrum teams, whether they be established or going through an Agile transition, is with the proactive management of their product and sprint backlogs.

Despite the proliferation of many great Agile / Scrum application life cycle management (ALM) tools, such as Jira, VersionOne, Microsoft Team Foundation Server, or others, teams struggle with the process of creating good backlogs, maintaining relevant information in backlogs, and regularly grooming them.

Based on observations, the problem arises from a combination of issues such as:

💣The product owner being overwhelmed or spread too thin
💣The lack of a clear product vision that translates into a product roadmap
💣The lack of a product roadmap that translates into clearly expressed release goals
💣Insufficient tools expertise or lack of understanding of Agile / Scrum best practices
💣Problems estimating the backlogs on time

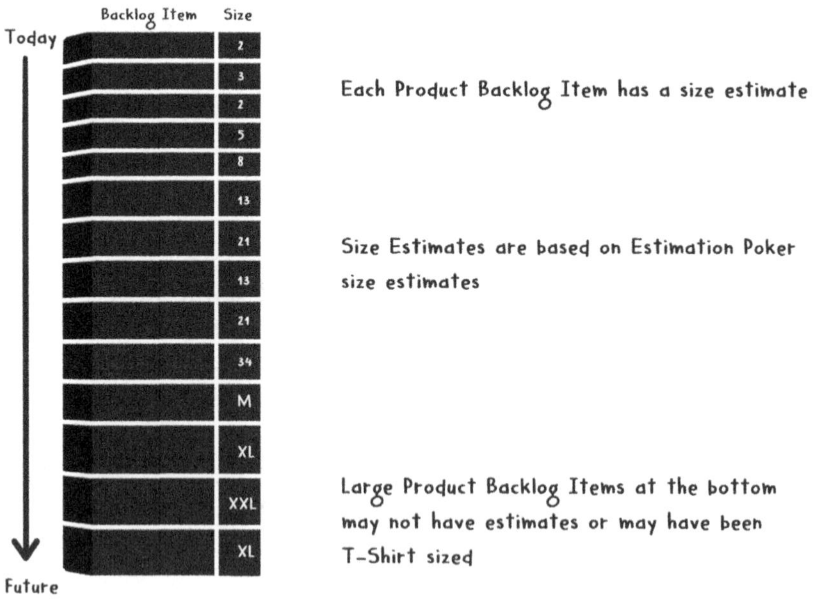

Figure 35 - Product Backlog Sizing

PREREQUISITES

To reiterate, some of the basic preconditions that have to be met in order to effectively manage the product and sprint backlogs are:

1. Good **user stories**
2. Good **estimation** techniques
3. Effective **release planning,** which is enabled by good user stories and good estimation techniques

Without good user stories and the application of good estimation techniques, the product and sprint backlogs become less useful and will ultimately prevent the Agile / Scrum process from becoming predictable.

In guiding the strategy and direction for the project, the product owner is responsible for providing the vision, product roadmap, release goals, and sprint goal. The product owner is expected to insert, re-prioritize, refine, or delete items from the product backlog; this can happen any time until the sprint scope is defined and committed to by the development team.

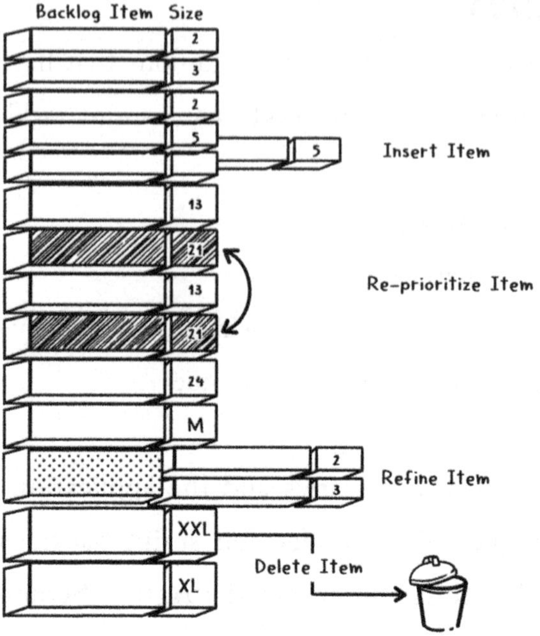

Figure 36 - Product Backlog Grooming / Refinement Process

Sizing and estimation of the product backlog usually occurs in sprint planning meetings or at regularly intervals during ongoing sprints. Depending on hAprow fast the product owner adds user stories, more frequent, maybe even daily, estimation sessions might be needed. The development team and product owner must work together to estimate the backlog items, that may be in sprint planning meetings or during ongoing regular sessions.

The important point is that product backlog items need to be estimated continuously in order to avoid the big bang estimation effort at one critical juncture. The frequency of the estimation effort is dependent upon the product / project development effort. This has the added benefit of exposing the entire team to the product owner's thinking, keeping everybody in sync and with an improved understanding of requested product features and functionality.

Sprint planning meetings or more frequent **estimation sessions** need to follow established Agile estimation best practices, meaning product backlog items are estimated using either estimation poker or T-shirt sizing.

The **product owner** is responsible for initiating requirements, creating user stories, and exerting scope control.

The **development team** is responsible for estimating the backlog in close collaboration with the product owner.

PRODUCT – RELEASE – SPRINT BACKLOGS

Many efforts require the management of three kinds of backlogs:

1. The **product backlog**, which essentially is the requirements database for the product, expressed in user story format – it exists for the lifetime of the product.
2. The **release backlog,** which is the subset of functionality that will have to be delivered in a specific release, according to the product roadmap and release plan – it is active for a specific release.
3. The **sprint backlog,** which represents the work to be done for the upcoming sprint – it is active for the sprint duration.

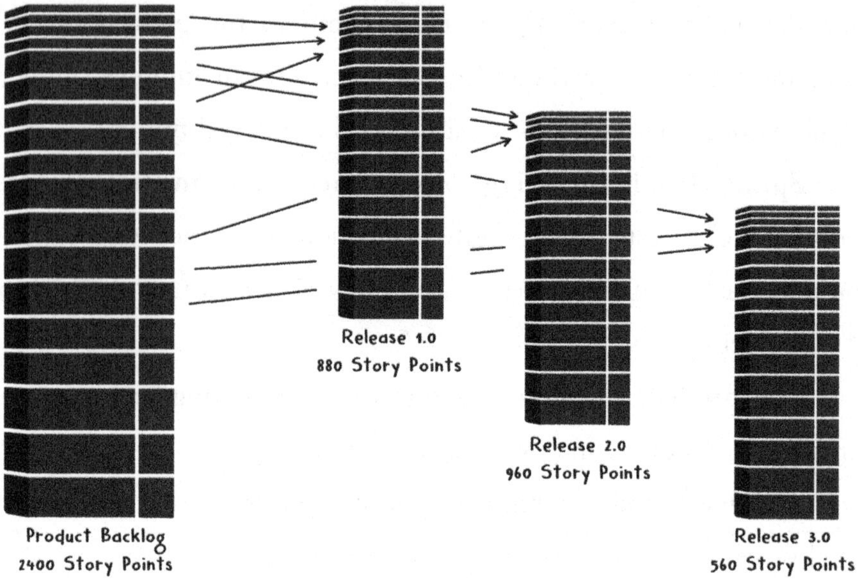

Figure 37 - Product Backlog feeding various Release Backlogs with Story Points

All tools in the market allow you go create tags or categories that allow for quick and easy views of which user story falls into which bucket.

Figure 38 – Sample Tags

Using such tagging lets you easily view relevant information about the specific user story and what state it is in.

Usually the next upcoming sprint is sized and estimated in detail using estimation poker, with the following two to three sprints and their contents being sized and estimated using either estimation poker or T-shirt sizing. Remember, once a sprint is committed to by the team, the scope cannot change, as you are "in-flight." Future sprints can still be adjusted and items can be reshuffled as agreed upon by the product owner and development team.

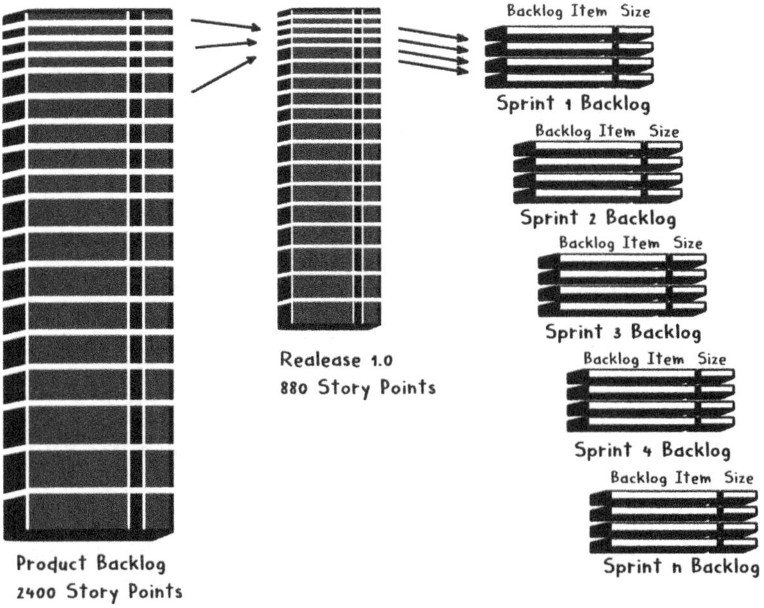

Figure 39 - Product Backlog to Sprint Backlog Refinement

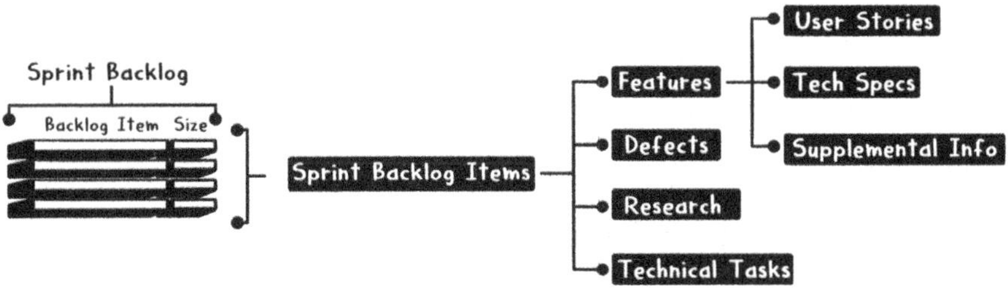

Figure 40 - The Sprint Backlog

Product backlog items (PBIs) and **sprint backlog items** may contain more than just user stories, but whatever is in the specific backlog is supposed to represent all work items that consume team capacity. This includes the actual user story describing the product feature to be implemented, defects, research, and other technical tasks.

Remember that the sprint backlog is supposed to be the most precise and fleshed out backlog due to its immediacy for actual delivery of a working product increment:

- Consists of committed PBIs negotiated between the team and the product owner during the sprint planning meeting
- Scope commitment is fixed during sprint execution
- Initial tasks are identified by the team during the sprint planning meeting
- The team will discover additional tasks needed to meet the fixed scope commitment during sprint execution
- Visible to the team
- Referenced during the daily scrum meeting
- Used to measure sprint acceptance by the product owner based on the definition of done

Backlogs are the Fulcrum of Agile / Scrum.

"Garbage in, garbage out" – in the case of Agile / Scrum backlogs, it is true.

The single most important artifact impacting the delivery success of the current sprint is the sprint backlog.

The single most important artifact impacting the success or failure of your product / project is your product backlog.

Agile / Scrum is a simple process, with few defined roles, few artifacts, and few events.

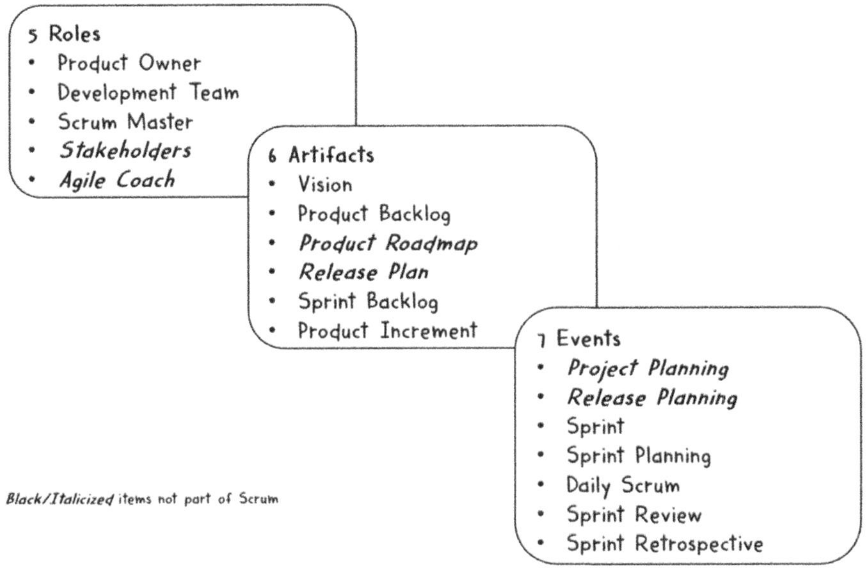

Figure 41 - Agile / Scrum Roles, Artifacts, Events

Backlogs hold the essence of what needs to get implemented, and as such need to be created, maintained, and refined with great care.

Unless the product owner and the development team focus on keeping backlogs up-to-date and well groomed, the Agile / Scrum process will not be able to deliver what is most important to organizations: **Predictability**.

COMMON ISSUES AND CHALLENGES

The section addresses some common issues and challenges that make managing product and sprint backlogs more difficult than it has to be.

Poorly Written User Stories

User stories should follow the **INVEST** principle. They should be:

- **Independent** (to the degree possible, user stories should not rely on other user stories to be implemented)
- **Negotiable**
- **Valuable** (feature focused, not task oriented, and are written in the user's language to the extent possible; where domain-specific language is required, such as in genomics or medicine, a subject matter expert will have to be available)
- **Estimable**
- **Small** (half a sprint or less for one development team member)
- **Testable** (must have acceptance criteria and not be subjective)

The testability of a user story is usually documented in the form of acceptance criteria on the reverse side of the index card (or the appropriate place in the ALM tool of your choice) and lists pass/fail testable conditions that help verify if the user story is implemented as intended.

All of this sounds simple enough, but frequently teams and product owners really struggle to write good user stories, and 90% of all problems encountered are based on user stories being too big.

When a user story is too big, it is hard to understand, to estimate, to implement, and is prone to wildly different opinions. So what is a good size for a user story? Basic guidance is that the user stories at the top of the product backlog (the ones that should have sufficient specificity in order to be worked on by a developer) should be sized such that a team member could easily complete two user stories during a sprint.

Good user stories are the basis for good backlogs. Please review the chapter "Quick & Dirty Guide to Writing Good User Stories" for more details.

Rapid Requirements Inflow

Sometimes the product owner might start adding user stories to the backlog far faster than the development team can estimate the effort. Then, out of desperation because the development team is busy with implementation tasks, the product owner might take on a solo effort of personally estimating stories without the development team being involved. This is a bad idea!

One of the core Agile principles is that "business people and developers must work together daily"; as such, the product owner cannot simply hand over user stories to the development team, as it defeats the purpose of jointly understanding and estimating user stories. One of the key characteristics of a good product owner is the ability to explain, in detail, what is meant by a user story.

If the inflow of requirements outpaces the team's ability to understand the requirements in detail and estimate them accordingly, the best course of action is to set up additional backlog refinement sessions to get things back on course.

Fantasy Wish List

The backlogs need to contain user stories that are relevant to the product, the release, and the sprints. It is counterproductive to clutter the backlogs with fantasy requirements that do not directly contribute to the product, release, or sprint goals.

Such wish list items need to be kept separate and managed outside of the backlogs, until such requirements become tangible goals at the product, release, or sprint level.

Backlogs = Requirements Specifications

Especially during Agile transformation / adoption efforts, teams often deal with a long tail of old artifacts available from previous software development methodologies.

For example, you might have many UML diagrams, functional or technical requirements specifications, etc. Why not put them to good use and just make them part of the user stories?

The reason not to do this is that specification has the side effect of shutting down discussions, as in "this is the specification we need to implement, nothing more, nothing less" – specification hinders the discover of newly emerging requirements.

However, this is not to say that once a user story is well understood by the team, you will never, ever, write a technical specification document in order to document the implementation. Write as much technical documentation as necessary to effectively aid your development process – but do not be a slave to past process artifacts.

Unavailable Product Owner

One of the most common challenges with product owners is their lack of availability to work with the development team. Product owners are enablers, making the development team more productive – or less!

Without an available product owner, the development team will get stuck – this might manifest itself either in sprint goals not being met, in poor quality, or in low velocity.

Lack of Backlog Grooming / Backlog Refinement

There really is no excuse not to regularly groom and estimate your backlog. As mentioned before, if for whatever reason your backlog is not being groomed, the best course of action is to set up additional backlog grooming sessions to get things back on course.

Chapter 9
Delivering Good Product Increments

Frequently, people in software development voice their frustration about things "still not working" or "not having improved" or "are being delivered with poor quality." These "things" usually refer to the software that is being delivered by their Agile / Scrum team – the **product increment**.

Within the context of a single sprint, a product increment is a work product that is the result of a sprint and is deemed "shippable" or "deployable," meaning it is developed, tested, integrated, and documented. A product increment may or may not be released externally, depending on the release plan.

Especially after an Agile transformation effort, frustrations abound because expectations for the transformation effort had been set too high. After all, the Agile transformation was supposed to solve all problems!

This chapter tries to address how to deliver good product increments for software. Although Agile / Scrum can be used for many other, non-software-related efforts, the author's expertise is centered on software product development, and this is what this chapter will specifically address. However, readers working in other domains might find it insightful as well.

THE IMPORTANCE OF FUNDAMENTALS OR BASIC PRINCIPLES

Too many organizations believe that once they have trained their employees in Agile / Scrum, all problems experienced in the past will disappear. That perception is misleading. One has to understand the underlying importance of the fundamentals or basic principles of software engineering to really make things work in Agile / Scrum and be able to deliver working, successful, good quality product increments.

Consider this analogy: You are given a watermelon and told it tastes sweet and juicy, and you should try it (somebody just sold you on Agile / Scrum!). So you take this watermelon and look at it. It is big, heavy, smooth, and tough skinned. It is far too big to stick in your mouth (you know how difficult it will be to make your organization truly adopt Agile / Scrum!), so you lick the skin (do a pilot project!). You surely disagree that the watermelon tastes good (the pilot project failed!).

If you take the superficial view, then the watermelon is tasteless and unpleasant (Agile / Scrum does not work!). Is that the truth? Of course not. You need to dig a little deeper to find the true value and taste of the watermelon. Once you get through the tough outer layers the watermelon is surely sweet and juicy.

It is the same with Agile / Scrum. If you look purely at the hype, you can easily be fooled. It looks spectacular – finally you will be Agile after all!

You must look beyond the flashy gimmicks to get to the truth. If you understand the fundamentals, the principles, the basics of software engineering, you cannot be fooled by the outward display and oversimplification presented to you.

HOW TO DELIVER GOOD PRODUCT INCREMENTS

As with any other complex discipline, what is essential to give you a chance of success is focusing on the fundamental basics. The more solid your basics are, the more likely you are to succeed.

For software development in an Agile / Scrum context, these basics break down into:

- Having the business side represented by least one knowledgeable domain expert who will act as the team's product owner.
- Having the product owner express requirements in the form of good user stories that are maintained in a product backlog /

release backlog / sprint backlog.

- Having the product owner and the development team work together to estimate the implementation effort.
- Implementing the software using a set of engineering best practices and DevOps best practices.
- Following and adhering to standard Agile / Scrum process, roles, artifacts, and events.

THREE MAIN FOCUS AREAS

Software projects still fail at an alarming rate. The reasons are many, but at the core of every bug that slipped through the process, source code that was lost, features that got implemented without corresponding requirements / user stories, or late effort integration efforts that turn out to be a grinding death march lies the insufficient use of fundamental best practices.

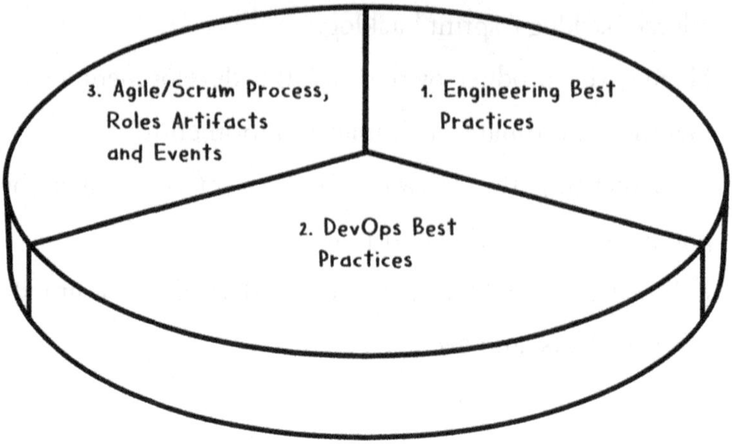

Figure 42 - Fundamental Best Practices

The three focus areas that determine a teams' ability to deliver good product increments are:

1. **Engineering** best practices
2. **DevOps** best practices
3. **Agile / Scrum** process, roles, artifacts, and events

Let us review each in turn.

Engineering Best Practices

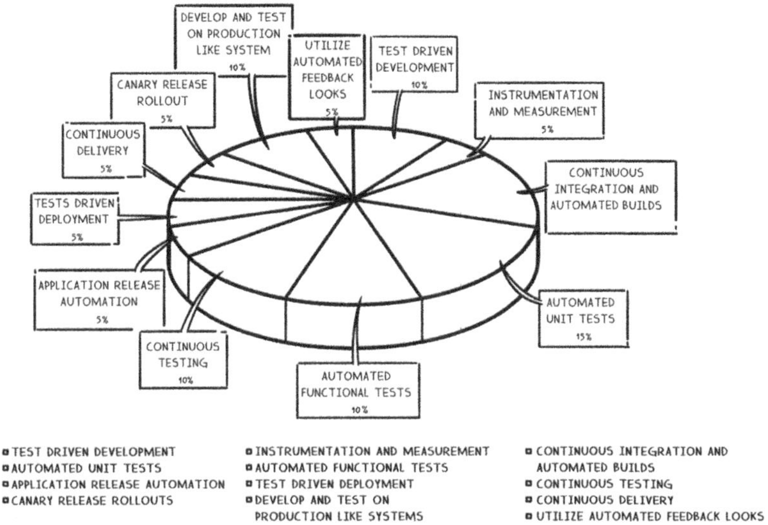

Figure 43 - Engineering Best Practices

Develop Iteratively

Most modern software development projects use some kind of spiral-based methodology rather than a waterfall process. There is a myriad of choices out there, from older legacy models such as the Rational Unified Process (RUP), to the last Agile / Scrum, and Extreme Programming (XP) frameworks. Developing iteratively follows the long established Deming Cycle (Plan-Do-Check-Act). Having a process is better than not having one at all, no matter what process you are following. In general, the process that is used is less important than how well it is executed and that it allows for continuous process improvement.

Establish Change / Requirements / Configuration Management

If you want to build good software, you need to express what you want to build, you need to track it, and you need to manage it by versioning it. Agile / Scrum focuses on user stories for expressing requirements; other methodologies go into various levels of depth and details using other formats (RUP utilizes use cases and UML, for example). Configuration management involves knowing the state of all artifacts that make up your system or project, managing the state of those artifacts, and releasing distinct versions of a system. You need to know what you build, with what version of the code, on what version of the environment.

Use Component Architectures and Design Patterns

Depending on your technology stack, you must pick and choose the appropriate architecture for your application. Component technologies, web services, and commonly used design patterns all provide best practice guidance for your application – useful knowledge of what works, what does not work, and why. Most importantly, these standards and patterns help you to not reinvent the wheel but to stand on the shoulders of others before you.

Use Source Control

Disregarding the often emotionally charged discussion about tool preferences, let me state that from this author's perspective, it does not really matter if a team uses Git, SVN, CVS, Team Foundation Server, or any other source control system – as long as you use it effectively. All of these systems have advantages and disadvantages, but in the end they are very much like choosing an Uber to get you to the airport – some are nice, some are big, some are small, but they all get you there.

Implement Unit Testing

A best practice for constructing code includes the daily build and smoke test. Martin Fowler[19] goes further and proposes continuous integration that also integrates the concept of unit tests and self-testing code. Kent Beck[20] supposedly said that "daily builds are for wimps," meaning only continuous integration would provide the best benefits from a coding perspective. Ultimately the frequency needs to be decided by the team, but the more frequent, the better. The reason for this is simple: It provides instantaneous feedback to the developer and allows for rapid problem resolution. Even though continuous integration and unit tests have gained popularity through XP, you can use these best practices on all types of projects. Standard frameworks to automate builds and testing, such as Jenkins, Hudson, Ant, and xUnit, are freely available.

Conduct Peer Reviews, Code Reviews, and Inspections

Make sure you know what is in your source code. It is important to

---

[19] https://martinfowler.com/

[20] https://www.amazon.com/Kent-Beck/e/B000APC0EY

review other people's work. Experience has shown that problems are eliminated earlier this way and reviews are more effective than testing – and cheaper on a cost per defect basis. Reviews and inspections also are effective as a learning / teaching tool for your peers.

Perform Testing

Testing is not an afterthought or an opportunity to cut back when the schedule gets tight. Disciplined testing in conjunction with extensive unit testing and reviews guarantees that software works. It is an integral part of software development that needs to be planned and executed. Platform proliferation (various operating system versions, various browsers, desktop vs. mobile platforms, etc.) has made testing critical to any project success. Testing still requires a good amount of manual effort, but compatibility testing across various platform combinations can be effectively automated using tools such as Selenium.

Automatically Measure the Process

How many developers are committing code every day? How many lines of code or class files? How many unit tests are executed per continuous integration run per day? As the project is coded and tested, the defect arrival and fix rate can help measure the maturity of the code. It is important to use a defect tracking system that is linked to the source control management system and the automated build system. By using defect tracking, it is possible to gauge when a project is ready to release by correlating information about the volatility of the code base to the defects found / fixed based on the set of tests executed. This kind of measurement has to occur automatically. Especially with continuous integration and automated platform tests, generating metrics needs to be automated due to the high frequency and large data volume that gets generated.

Use Smart Tools

There are more and more effective tools that help software development teams address all kinds of problems, from coding standards, to security scans. Utilize as many of the static[21] and dynamic[22] code analysis tools as you can in as much of an automated way as you can to make your life easier.

---

[21] https://en.wikipedia.org/wiki/List_of_tools_for_static_code_analysis

[22] https://en.wikipedia.org/wiki/Dynamic_program_analysis

DevOps Best Practices

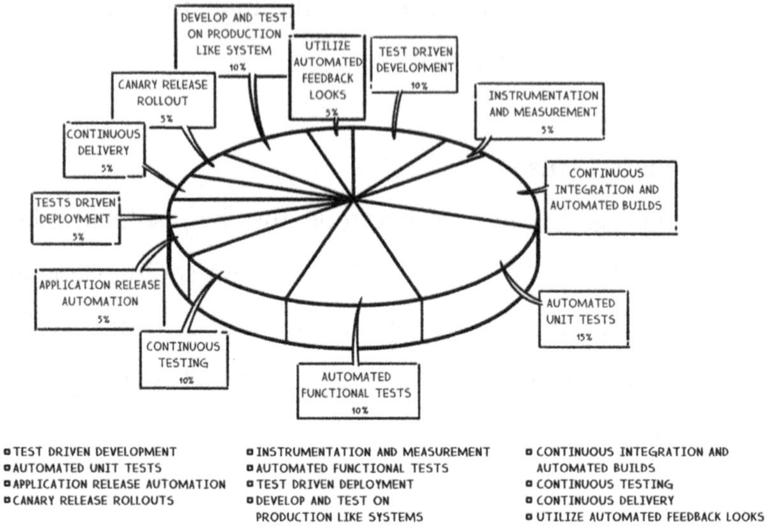

Figure 44 - DevOps Best Practices

Test-Driven Development, or at a minimum the implementation of a disciplined unit testing approach – Test-driven development (TDD) is a software development process that relies on the repetition of a very short development cycle where requirements are turned into very specific test cases, then the software is improved to pass the new tests. This is opposed to software development that allows software to be added that has not proven to meet requirements.

The key point here is that each requirement is developed based on test criteria. Whether your project follows this strict TDD approach, or if you simply ensure that your software has corresponding automated unit tests for every software change, does not matter, as long as you implement tests for every change. It is hard to stress how much this approach improves quality, system stability, and maintainability – it is really the basis of building a quality product from the ground up – especially for new projects.

But, what to do if the project has been active for 8 years and has one million lines of code but no automated unit tests? Do as the boy scouts do – whenever and wherever you touch code, make it better by adding unit tests. Every time you add code, add unit tests. Every time you refactor a piece of code, create unit tests. That way you will slowly transform your code base.

Instrumentation, across the product or solution, including infrastructure – Instrumentation at all levels is critical for keeping a handle on your software, your infrastructure, and ultimately your customer satisfaction. Every system built is different, and every specific domain has its own challenges, but simply put, you should be able to tell at any given time:

- How many users are on your system?
- What their activity level is.
- The number of transactions that succeeded / failed.
- The utilization numbers for various server side health indicators, like memory utilization, database connections, uptime, disk space utilization, etc.

Continuous Integration

Continuous integration (CI) is a development practice that requires developers to integrate code into a shared repository several times a day. Each check-in is then verified by an automated build, allowing teams to detect problems early. CI allows for 3,000-mile-high monitoring of the development team – especially important in today's distributed development model – such as understanding who checks in code how frequently, who breaks the build, how many tests fail, how many tests are added, etc.

Automated Builds

Build automation is the process of automating the creation of a software build and the associated processes, including: compiling computer source code into binary code, packaging binary code, and running automated tests. Automated builds are usually a separate task / activity from CI, because automated builds focus on packaging the software solution into something consumable for a target test environment (for example by automatically deploying to a set of virtual machines for testing purposes), whereas CI only tries to accomplish its tasks within the CI environment.

Automated Testing

Automated testing tools are capable of executing tests, reporting outcomes, and comparing results with earlier test runs. Tests carried out with these tools can be run repeatedly, at any time of day. The method or process being used to implement automation is called a test automation framework.

Continuous Testing

Continuous testing is the process of executing automated tests as part of the software delivery pipeline to obtain immediate feedback on the business risks associated with a software release candidate.

Application Release Automation

Application release automation refers to the process of packaging and deploying an application or update of an application from development, across various environments, and ultimately to production.

Test-Driven Deployment

Test-driven deployment refers to the approach of automatically validating, via automated test scripts, that the deployment of a build or release to the target environment was successful. This can range from simple smoke tests on the target environment, to full, detailed, end-to-end tests that validate that the deployed functionality is working as expected.

Continuous Delivery

Continuous delivery (CD) is an approach in which teams produce software in short cycles, ensuring that the software can be reliably released at any time. It aims at building, testing, and releasing software faster and more frequently. The approach helps reduce the cost, time, and risk of delivering changes by allowing for more incremental updates to applications in production. A straightforward and repeatable deployment process is important for continuous delivery.

Canary Release Rollouts

Canary releases are designed to reduce the risk of introducing a new software version in production by slowly rolling out the change to a small subset of users before rolling it out to the entire infrastructure and making it available to everybody. A good example of this is how Android app publishers can configure their app for Google Play Store Beta distribution accordingly, but the approach can be done regardless of the technology used. As a matter of fact, the author used this very approach some 25 years ago when shipping an accounting software package to some 30,000 users – the company would ship the new software (on CDs) to 1,000 users and wait for feedback. If there were no complaints, the next 2,000 copies would be sent out, wait again, and then finally release to the remaining user base.

Another approach used today is the so called Blue/Green/Deployment approach (sometimes called A/B/Deployment), where you have two similar production environments, one with the current version (Blue/A), one with the future version (Green/B), and you switch users from one environment to the other without their knowledge. Again, regardless of what approach is used, the purpose is to make the deployment of software release transparent to the end users and minimize risk to the company.

Develop and Test Against Production-like Systems

This concept stresses the importance of developing and testing on production-like systems, down to the specific operating system patch level and supported libraries. Also, if you are dealing with localization / internationalization projects, make sure to run your application / app on the localized system just as your end users will.

Utilize Automated Feedback Loops

The last obvious best practice is automated feedback loops – and it is closely related to the aforementioned Instrumentation. It is important to stress the need for automation here again – this cannot be a manual effort; it has to be automated. With all the approaches / processes mentioned above, many of them automated, there is simply no way for a human to keep up. Feedback loops have to run constantly, update dashboards, and alert the right folks at critical junctures. For example, if a system runs "hot" and runs out of memory, an alert has to go out to your DevOps engineer to review what is going on; that DevOps engineer might in turn pull in developers to see what is wrong. The automation might even take down the server and bring another server online. Unless this is automated, there is no way the feedback loop will be effective – it will be too slow, and in the above example, your hot running system will crash.

Agile / Scrum Process, Roles, Artifacts, and Events

Engineering best practices and DevOps best practices are "embedded" in the Agile / Scrum process.

This chapter is not about explaining Agile / Scrum from scratch, but as a reminder, here is the process in a nutshell:

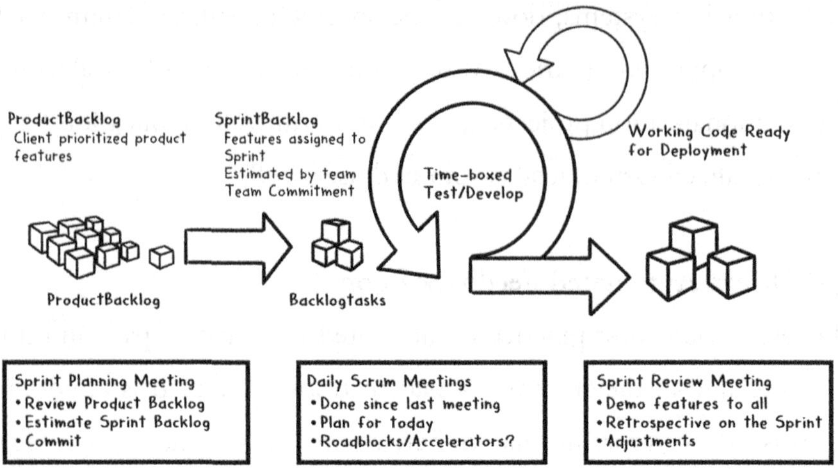

Figure 45 - Agile / Scrum Process

The **Agile / Scrum process** has a set of defined roles, artifacts, and events. Again, without getting into the details, here is a quick overview:

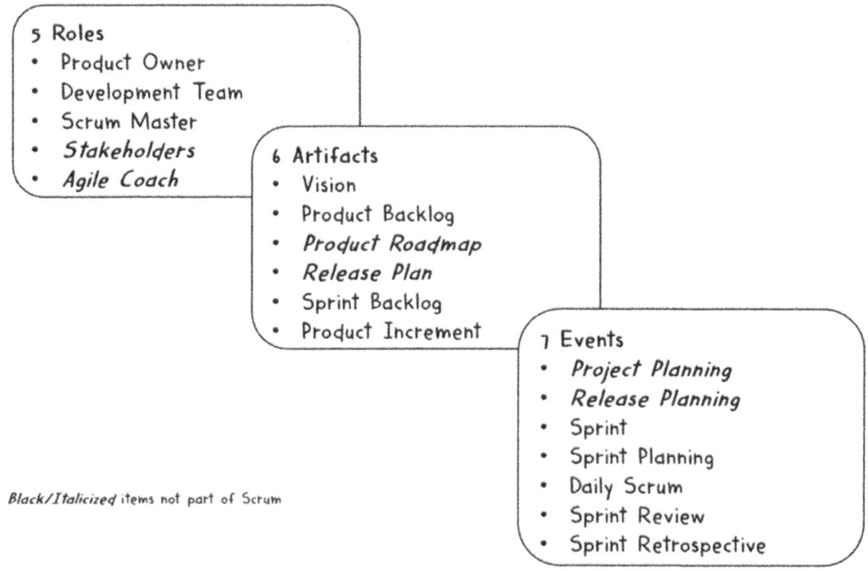

Figure 46 - Agile / Scrum Roles, Artifacts, Events

The main point to stress here is that in order to deliver good product increments, you need to have a well-understood process, with prescribed roles, artifacts, and events – that is what provides the repeatability.

Best practices in turn provide the qualitative aspect of the product increment and in many ways provide the framework for scaling up. With very few users, you might get away with not following some (or even all) of the best practices mentioned, but with millions of users, your best practice adherence becomes obligatory.

WHAT IF YOU CANNOT DO IT ALL?

Another disclaimer here: This is the author's totally ***subjective*** list of top 9 practices / process areas to focus on if you cannot do them all.

The list is presented in order of importance or impact. Again, it is a subjective list, based on industry experience, with a brief justification for each.

1. **Good Development Team**. A good development team can compensate for a lot, including missing roles or bad practices, so having the right people on the development team is crucial. But remember, a team is more than a bunch of "A" players; they need to work effectively together.
2. **Good Product Owner**. A good product owner who is able to succinctly express in user stories what is needed for the product, and who has the right personality to work through issues with the team, is crucial. In many ways, the product owner is like the team's quarterback.
3. **Good Sprint Backlog - User Stories**. You must know what you need to build to be successful. The immediate concern is the sprint backlog and good quality user stories. Good product owners feed good backlogs.
4. **Good Product Backlog - User Stories**. For mid- to long-term success, you need a good product backlog. Again, good product owners feed good backlogs.
5. **Good Process - Sprint Planning**. Of all the process items you might think are important, sprint planning stands out. Performing disciplined sprint planning, based on good user stories out of the product backlog, using solid estimation techniques (estimation poker or T-shirt sizing), is pivotal to

delivering good product increments at a predictable velocity.

6. **Use of Component Architecture and Design Patterns.** Not reinventing the wheel, and relying on industry standards, will make sure you do not face big technical faux pas.
7. **Develop Iteratively / Small Batches.** Incremental development in small batches keeps everybody honest because you deliver all the time. It shows real progress, and is tangible. It also helps the team establish aforementioned best practices more easily than in big bang development effort.
8. **Continuous Integration and Automated Builds.** Using continuous integration and building all the time validates that you can pass the first hurdle, successful compilation, and put the build together. This, in turn, establishes system stability, which is crucial for downstream activities such as testing.
9. **Automated Unit Tests.** Last but not least, automated unit tests help catch problems way up front in the cycle, avoiding the dreaded late discovery of bugs. Automated unit testing is a big bang for a modest buck, helping on all fronts, from improving developer productivity, gaining confidence that a fix did not break anything unforeseen, to building quality from the ground up. A big win / win for the team and the product.

THINGS KEEP CHANGING – STAY FOCUSED ON BASICS

Yesterday's best practice may not be one today, and you may not know what best practices might emerge tomorrow. Beware of new acronyms and hype cycles.

For example, the xUnit framework has been around since the mid-1990s (Kent Beck's first implementation was Smalltalk based, and ported over to other languages in subsequent years), but only gained industry-wide support in the 2000s. Now it is by far the standard when it comes to automated unit testing frameworks – for any implementation language. Who knows what will be used in 10 years?

Ten or 15 years ago, virtual machines revolutionized data center operations. Today, it is all about containers, Kubernetes, Mesosphere, and Docker Swarm.

Fifteen years ago, Agile / Scrum was a major differentiator for companies trying to get to market faster than their competition. Many considered it a competitive advantage. Today, it is considered a necessity - if you do not follow an Agile / Scrum process, you are considered to be putting your business at risk.

Things change, best practices change, but many of the fundamental basics still apply. Stay vigilant and open minded to change, take the new things that are good, and ignore the hype items, but above all, remember the basics!

CHAPTER 10
AGILE ESTIMATION HOW TO

One of the key areas that sets traditional process frameworks, such as Waterfall or V-Model, apart from Agile / Scrum is estimation techniques and approaches.

Confusion often surrounds Agile / Scrum terms such as **story points, estimation poker, T-shirt sizing, velocity** – and how all those terms relate to better and more precise estimates.

This chapter will quickly summarize the key issues addressed in Agile / Scrum that enable teams to better and more reliably estimate their development efforts.

ESTIMATION UNCERTAINTY

It is worth noting that the meaning of "estimate" can be

1. to roughly calculate or judge the value, number, quantity, or extent of something [verb], and
2. an approximate calculation or judgment of the value, number, quantity, or extent of something [noun]

Synonyms include: *calculate roughly, approximate, rough guess* – hence the slang term "guestimate" – a mixture of guesswork and calculation.

Dwight D. Eisenhower famously said: *"Plans are worthless, but planning is everything. There is a very great distinction because when you are planning for an emergency you must start with this one thing: the very definition of "emergency" is that it is unexpected, therefore it is not going to happen the way you are planning."*[23]

Or, as Mike Tyson put it more succinctly: "Everybody has a plan until they get punched in the mouth."

The point is that estimates are never precise, but in project management – especially software / IT project management – estimates regularly are baselined, cast in stone, tracked against, and then cause many political problems because it is perceived that the people that originally estimated the effort got it all wrong.

They did not get it wrong; they just did not know enough at the time.

---

[23] From a speech to the National Defense Executive Reserve Conference in Washington, D.C. (November 14, 1957) ; in Public Papers of the Presidents of the United States, Dwight D. Eisenhower, 1957, National Archives and Records Service, Government Printing Office, p. 818 : ISBN 0160588510, 9780160588518

CONE OF UNCERTAINTY

The **cone of uncertainty** is a project management term used to describe the level of uncertainty existing at different stages of a project.

Whenever you estimate an effort at the very beginning, say at the concept phase of an idea, you are likely to be magnitudes off in terms of precision. As you go through different project phases and you learn more about the actual work and challenges involved, the better (meaning the more closely approximated to the actual work effort) your estimates will become.

In short, we become more certain of our estimates as we learn more about what we are estimating.

Estimation variability decreases the closer we get to project completion.

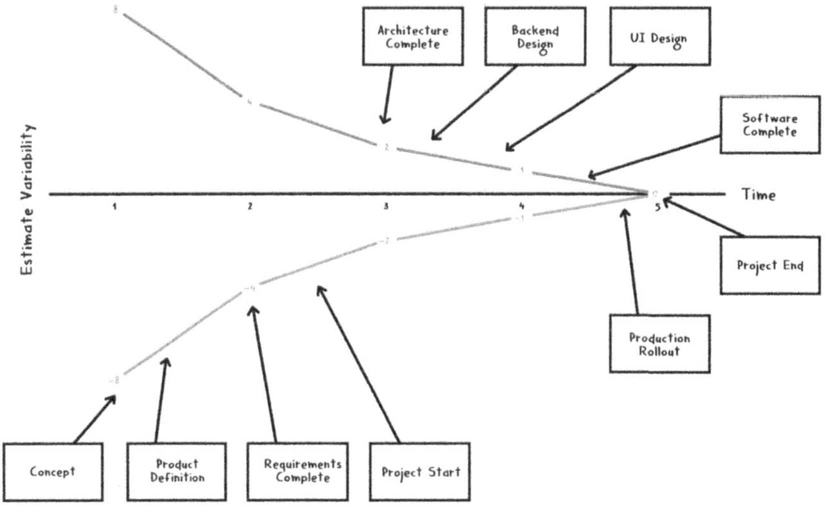

Figure 47 - Cone of Uncertainty - Estimation Variability

The only exception to this is if you are working on a process that is 100% understood. A good example is a production line; you know exactly how long the piece will move through different stages on the production line, and you can estimate exactly how long it will take to finish one instance of your product.

Production lines usually have been optimized with lots of empirical time measurements, ensuring that each production step runs smoothly.

If you are doing anything that involves uncertainty, creativity, or something that has never been done before, the cone of uncertainty rules. This applies to most software development and IT projects.

Even regular IT tasks such as server upgrades, frequently take longer than expected because they "get punched in the mouth" – maybe your system administrator runs out of disk space, experiences an unexpected power outage that corrupts the upgrade, or faces some other "emergency."

You can automate the process and make it more predictable, such as Amazon Web Services automated server requisitions in the cloud. However, this usually requires process standardization, which basically means it is not something that involves uncertainty or creativity or has never been done before – to the contrary, it is something that is known, well understood, and repeatable.

LONG- AND SHORT-TERM PLANNING HORIZONS

Decreasing estimation variability is dependent upon the **planning** horizon.

If you plan what you are going to do today, you most likely will have a good understanding of what your day looks like, and therefore can estimate with high probably (or low estimate variability) that you are going to make it to an appointment at 5:00 PM this afternoon.

On the other hand, if you are trying to plan for something 5 months out, the best you can do is put down a reminder – the chances of something unforeseen happening that might change your plans are high. You might get sick, be out of town for unexpected business travel, or need to tend to your elderly parents.

For estimation of software projects, this translates into the traditional "planning onion" as follows:

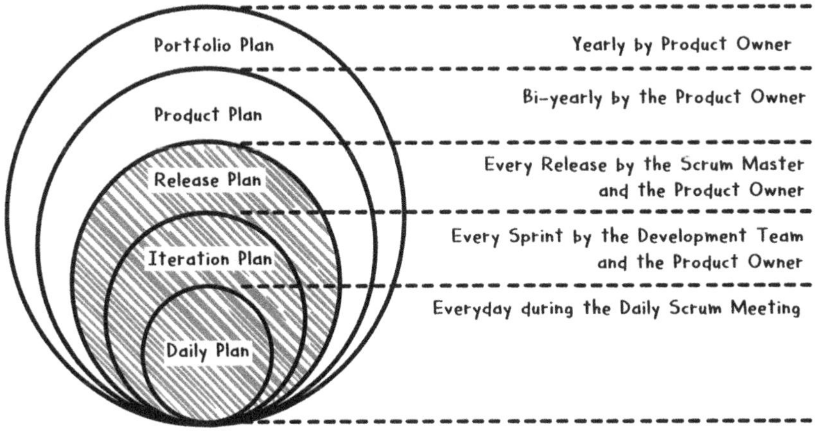

Figure 48 - Traditional Planning Onion

This planning onion basically works from the outside in to provide ever increasing levels of detail; as such it is closely related to the concept presented in the cone of uncertainty.

To reflect the timing aspect and estimation detail, you can think of the planning onion like this:

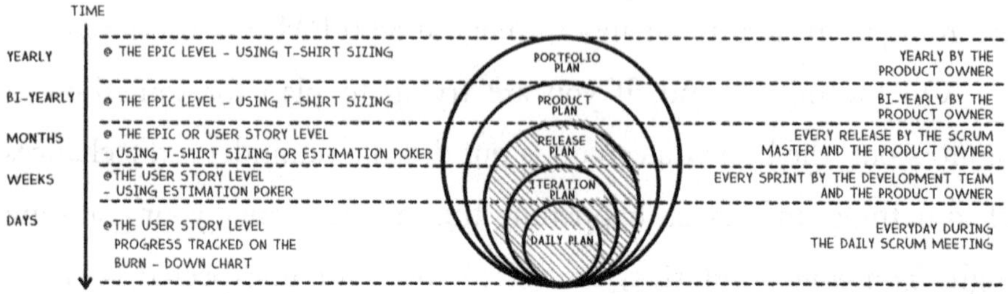

Figure 49 - Traditional Planning Onion Showing Timing and Planning / Estimation Levels

Finally, for Agile / Scrum estimation time horizons, you can think of long-term vs. short-term planning horizons as follows:

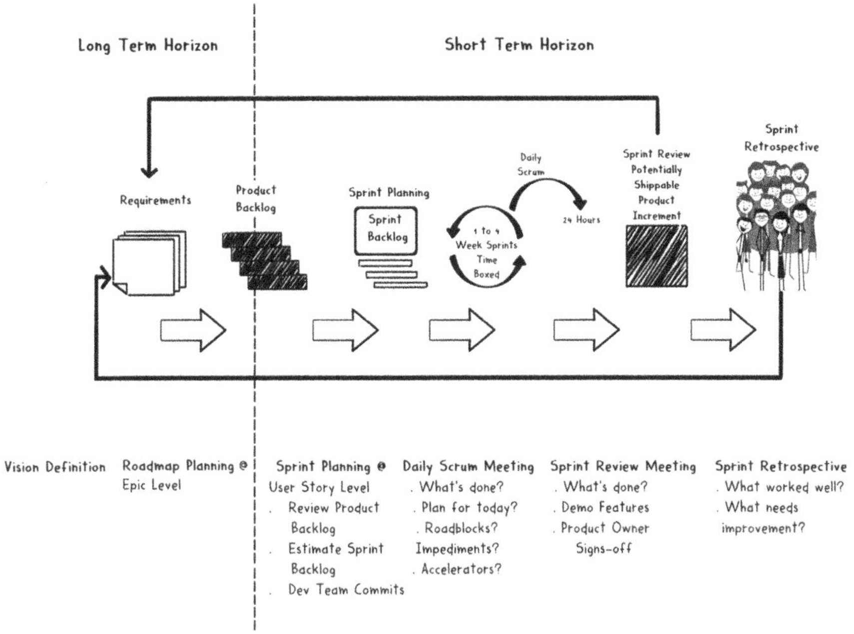

Figure 50 - Long and Short Term Planning Horizons

Vision and roadmap planning are long-term horizon activities, whereas sprint planning falls into the short-term horizon activities.

PLANNING AND ESTIMATION DIFFERENCES BETWEEN WATERFALL AND AGILE / SCRUM

So, what are the actual planning and estimation differences between Waterfall and Agile / Scrum?

From an estimation technique perspective, there is no reason why T-shirt sizing or estimation poker could not be used on either methodology.

The main problem with waterfall is that it attempts to esti-

mate the entire project upfront, thereby committing the project to an unrealistic timeline early on, without clearly understanding the scope.

The following picture shows a traditional waterfall process with three imaginary (but for IT / software projects very common) "interruptions," which basically reset the project to the beginning.

1. **Requirements change** – key requirements changed and force the team back to the Analysis drawing board.
2. **Customer management turnover** – your customer / client management or key stakeholders changed, and the "new team" wants to provide input (read "new direction") on the project, which causes a reset.
3. **Technology innovation** – new technologies allow for better ways to deliver the project, which forces the team to re-evaluate existing analysis, design, and code and start over

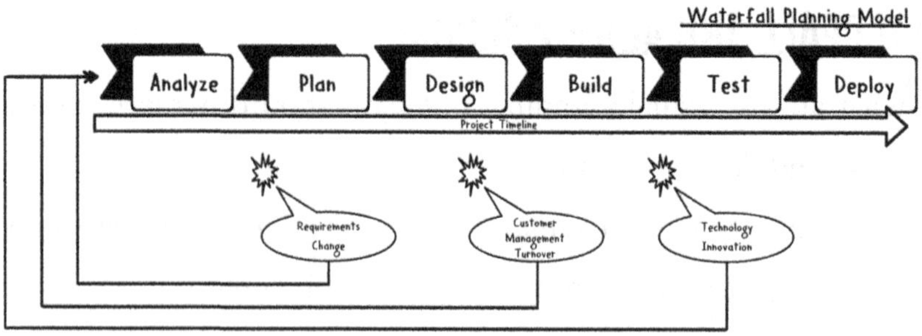

Figure 51 – Waterfall Planning Model

On waterfall projects, this happens all the time! You reset without delivering value.

Using Agile / Scrum, on the other hand, you are able to deliver small product increments rapidly. Your planning horizons are shorter, and the Agile / Scrum team basically commits to delivering a working increment at the end of every sprint.

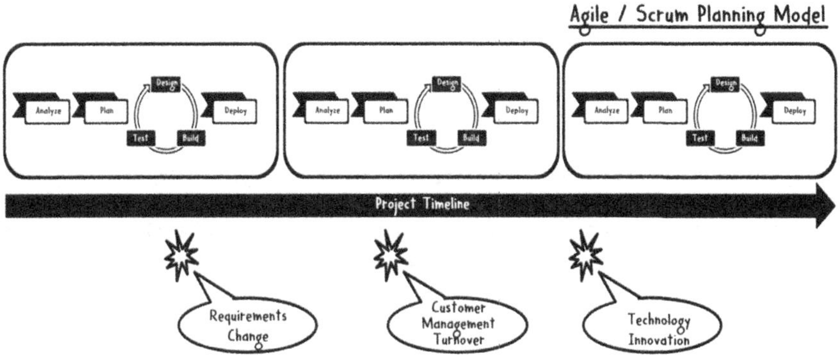

Figure 52 – Agile / Scrum Planning Model

Even if requirements change mid-sprint, the product increment at the end of that sprint will be delivered. New requirements make it into the product backlog to be prioritized for later sprints.

Even if customer / client management or stakeholders change, in Agile / Scrum the project team can demonstrate working software at the end of every sprint, many times lessening the desire to "change direction" (because, after all, it works already!). Any changes requested are logged as additional requirements in the product backlog to be prioritized for later sprints.

Lastly, even technology innovation can be accommodated by including it into the product backlog to be prioritized for later sprints.

The important thing here is that in Agile / Scrum, despite interruptions, unexpected developments, and "emergencies," the team keeps delivering working software in the form of product increments at the end of every sprint – delivering real tangible value.

Contrast that with the waterfall methodology, where the only value created often is a large set of documents that the team struggles to keep updated.

Delivering working software also creates "facts on the ground" that are harder to ignore or debate than if the team had merely produced a set of documents.

COMMON ESTIMATION APPROACHES

What are some common estimation approaches and how do they work?

Absolute vs. Relative Estimation

One key thing to keep in mind is that Agile / Scrum teams use **relative estimation**. This is because it is relative to the team. What does that mean? Essentially it means that the measurement unit / granularity agreed upon for a task is specific to the development team estimating it. Team A could estimate it at 8 story points, and Team B might estimate 5 story points. It is relative to the team, and a relative estimate does not try to be 100% precise by using absolute measure.

The following picture tries to clarify the differences:

Figure 53 - Absolute vs. Relative Estimation

The **absolute** measure here is milliliters (ml), which is an International System of Units fluid measure. There is no interpretation as to what it means. It is absolute. Wine bottles are classified accordingly.

The **relative** measure might be the small, medium, large classification used in T-shirt sizing or the 3, 5, 8 story point Fibonacci number sequence used in estimation poker (more about estimation poker in the next section).

The following table compares and contrasts absolute vs. relative estimation techniques:

ABSOLUTE	RELATIVE
Hours / days	Story points / T-shirt sizes
Used for sprint planning	Used for release planning
Accuracy is important	Accuracy is not important
Does not eliminate bias	Eliminates bias
Supports comparison	Cannot be compared with another team's performance – each team is unique. Some teams might produce 35 story points per sprint; some might produce 60.

Table 7 - Absolute vs. Relative Estimation

Estimation Poker

Figure 54 - Estimation Poker Using Fibonacci Numbers

As mentioned before, **estimation poker** is a relative estimation technique that favors accuracy over precision. It commonly uses the Fibonacci number sequence (1, 2, 3, 5, 8, 13, 21, 34, 55, ...). By definition, the first two numbers in the Fibonacci sequence are 0 and 1, and each subsequent number is the sum of the previous two.

The development team members doing the work have to do the estimates. You cannot have people other than the actual development team estimate the work. This is critical in order to create accountability and make sure the development team is fully committed to the estimate.

Estimation poker self-corrects for team idiosyncrasies – meaning two teams might estimate the same work and come up with different estimates because the specific team's composition and skill set might be different.

The following 8 steps describe the basic estimation poker process:

Step 1 – Identify an "anchor story" – the best understood story everybody can agree on and provide a size / effort reference.

Step 2 – Product owner explains story.

Step 3 – Each development team member selects a card (secretly).

Step 4 – Together, all development team members show their cards.

Step 5 – If different, the high and low estimators briefly explain their choice and assumptions.

Step 6 – The product owner provides additional information, as necessary.

Step 7 – The development team iterates the process, up to 3 times, until consensus is reached.

Step 8 – Repeat for all stories that need to be estimated, always keeping the anchor story in mind.

T-Shirt Sizing and Affinity Estimation

Figure 55 - What Size?

T-shirt sizing (aka affinity estimation) is based on the need to estimate things quickly without necessarily having all the details available that you usually might require. Here are some examples of how T-shirt sizing can be used:

1. For budgeting purposes – when a vision and roadmap are being discussed, T-shirt sizing can do the trick by having some experienced developers look at the high-level tasks and sizing them appropriately. The same applies for staffing plans (how many developers do we need to hire next year?).
2. When user stories are not available – product owners are the detail level domain experts who know what their product is supposed to do in detail. During estimation poker, the product owner explains, discusses, and clarifies each user story. But the reality is that oftentimes fully fleshed out user stories might not be available yet, nor will the product owner sometimes know 6 months ahead of the potential sprint what necessarily has to be developed. Having said that, at a minimum you should have a placeholder subject line for each user story you wish to estimate, for example, "Support HP OfficeJet 9800" – at least developers will quickly be able to grasp what is required in the larger scheme of things without knowing the exact details.
3. When you have 2,986 user stories – it is not uncommon for a productive product owner to crank out hundreds and even thousands of user stories – at varying levels of detail, some fully fleshed out, and some only with a subject line – the challenge becomes the excessive amount of time it would take to

estimate all those stories. Do you put your development effort on hold for 6 weeks simply to generate estimates with estimation poker?

The answer to these estimation challenges is to use T-shirt sizing. In order to have a somewhat comparable scheme between estimation poker and T-shirt sizing, it is a good idea to use the following quantification scheme:

- XS = 1 story point
- Small = 2 story points
- Medium = 3 story points
- Large = 5 story points
- XL = 8 story points
- XXL = 13 story points
- EPIC = Product owner needs to break down the epic into more refined user stories. (For example, the epic might state something like "Internationalization Support." This needs to be broken down into more detailed user stories like "Provide Support for British English," "Provide Support for Spanish," in order to be able to estimate it.)

Then, follow these four easy steps:

Step 1 – Take 1 minute and have the development team agree on a single "small" user story. Repeat with XS, medium, large, XL, XXL, and EPIC sized stories.

Step 2 – Take 10 – 30 minutes and have the development team put all remaining user stories into categories of XS, small, medium, XL, XXL, and EPIC.

Step 3 – Take another 10 – 30 minutes and have the development team review and adjust user story categorizations as necessary – Done!

Step 4 – Have the product owner review the categorizations.

This 4-step process allows the development team to tackle many hundreds of user stories quickly.

A common challenge is that with a development team of 7 (+/-2), you will experience disagreements. Some people will argue that a user story is a 5, others will bet it is a 13. So, what to do?

COMMON RESOLUTION APPROACHES

There are basically 4 ways to resolve disagreement during the estimation process:

1. Ask the outliers to plead their case and try to convince the other development team members. Basically this is a negotiation back and forth until consensus is reached. This kind of open discussion works well with equally balanced team members but can be challenging with strong, Type A, personalities

and unbalanced teams, for example, a team consisting of 3 very strong technical engineers and 4 entry level engineers.

2. Do a silent vote or a "fist of five" consensus model.
3. Majority votes are popular because the process avoids conflict, but majority votes usually end up running the risk of creating a committee decision where nobody feels responsible – you want accountability, so this is not a good approach.
4. Choose the highest estimate, but this creates the risk of estimate bloat.

The best way to resolve disagreements is to have the discussion and get all team members to agree on the estimate.

Remember that estimation is team-specific, and that the estimate is basically the development team's commitment to deliver based on this estimate – so consensus is important.

All for one, and one for all!

Regardless if you are using estimation poker or T-shirt sizing, the cone of uncertainty still applies. Be aware that using T-shirt sizing early in the process will produce estimates with higher variability.

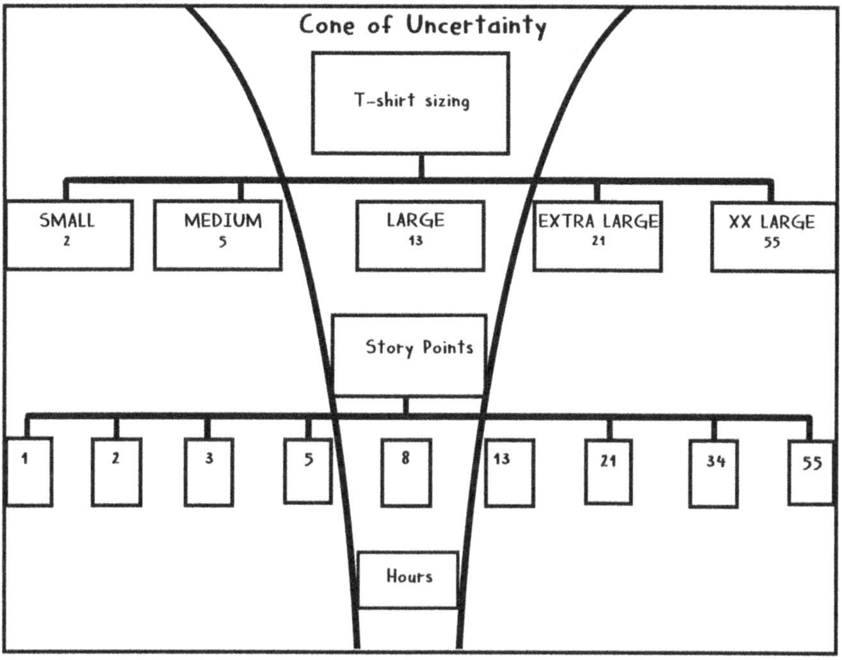

Figure 56 - The Cone of Uncertainty as Applied to Agile / Scrum Estimation and Planning

Depending on what you are challenged with estimating, choose the right estimation method:

- Roadmap planning => Use T-shirt sizing.
- Release planning => Use T-shirt sizing or estimation poker, depending on the number of user stories.
- Sprint planning => Use estimation poker.

ESTIMATION TEAM SIZE

The estimation team size is determined by the Agile / Scrum development team size, which should be 7 (+/-2).

The people who will do the work need to be the ones who estimate it.

Sometimes organizations think that adding more people to the estimation effort will make the estimate better, but that is a commonly accepted fallacy. Larger teams do not produce better estimates:

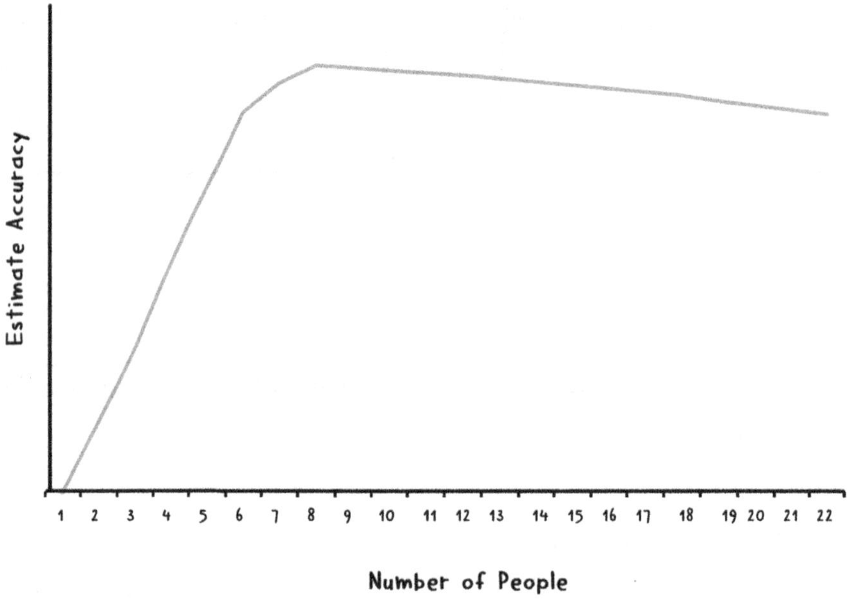

Figure 57 - More People Does Not Mean Better Estimates!

Estimation accuracy actually decreases with increasing team size.

Extensive research has been done on team size considerations. It is generally acknowledged that the optimal team size is somewhere between 5 and 8 team members. Some teams have been shown to function well with up to 12 team members. Co-location, offshoring, partnering, and staff qualification all affect the estimation efficiency.

It is worth pointing out that sports teams are organized around this size (soccer: 11; basketball: 5; baseball: 9; football: 11). The exception seems to be rugby, which has 15. Military squads are organized around this size as well.

For estimation efficiency as well as team dynamics, it is important to understand the following:

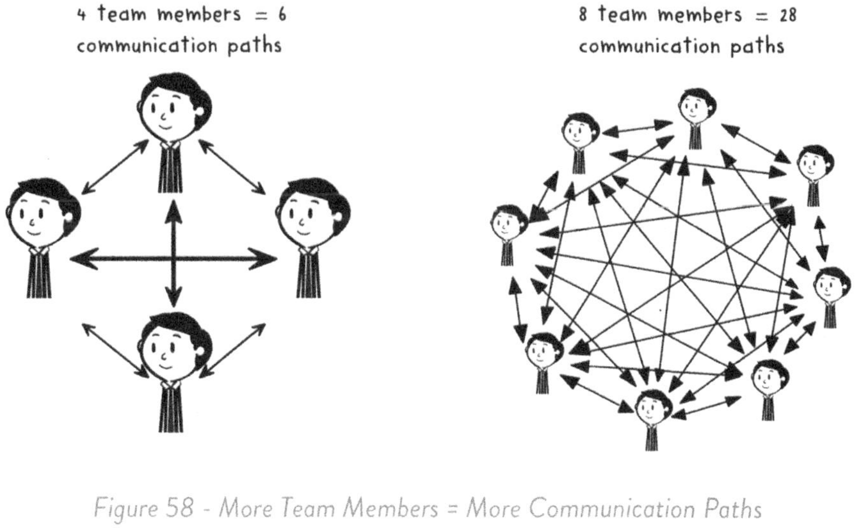

Figure 58 - More Team Members = More Communication Paths

The bigger the team, the harder it is to effectively communicate, estimate, and come to consensus. The bigger the team, the more likely something will fall through the cracks because of mis-

communication.

This challenge of ever increasing communication paths with larger teams is also somewhat related to the famous "Dunbar's number,"[24] which suggested a cognitive limit to the number of people with whom one can maintain stable social relationships (150).

WHAT IS VELOCITY?

Velocity is the rate at which the development team can reliably deliver story points (usually packaged into time-boxed sprints and resulting in a working product increment).

Definition	1. Rapidity of motion or operation; swiftness; speed: a high wind velocity 2. The time rate of change of position of a body in a specified direction 3. The rate of speed with which something happens; rapidity of action or reaction 4. Points delivered by a team per sprint
How to Calculate	Rolling average of sprints Max. Min. Lifetime average
How to Use	To determine sprint scope To calculate approximate cost of a release To track release progress
Remember	Velocity of two teams is not comparable Velocity changes when the team composition changes Velocity increases with team's tenure Velocity does not equal Productivity Defects and rejected stories count against velocity!

[24] https://en.wikipedia.org/wiki/Dunbar's_number

Estimates Self-Correct

When using estimation poker, estimation variability decreases usually over the course of the first 4 to 6 sprints. It is rare to see stable teams that do not achieve fairly accurate estimates after working together for 5 to 6 sprints.

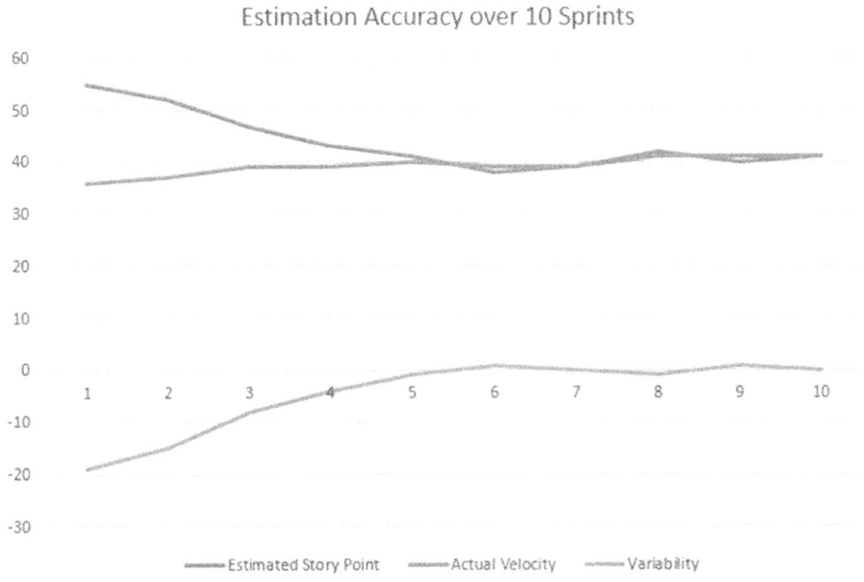

Figure 59 - Estimation Accuracy Improves over Time

Please note that "stable teams" is key here – teams need to be stable, meaning team members need to know each other, have worked with each other, and successfully formed a team – following the standard "forming–storming–norming–performing" model of group development, first proposed by Bruce Tuckman in 1965.[25]

---

[25] https://en.wikipedia.org/wiki/Tuckman's_stages_of_group_development

These phases are all necessary and inevitable in order for the team to grow, to face up to challenges, to tackle problems, to find solutions, to plan work, and to deliver results.

ESTIMATE ACCURACY AND VELOCITY OVER TIME

Once a development team has gone through the "forming–storming–norming–performing" process, team velocity is established. Estimation accuracy and predictable velocity allow for longer term forecasting.

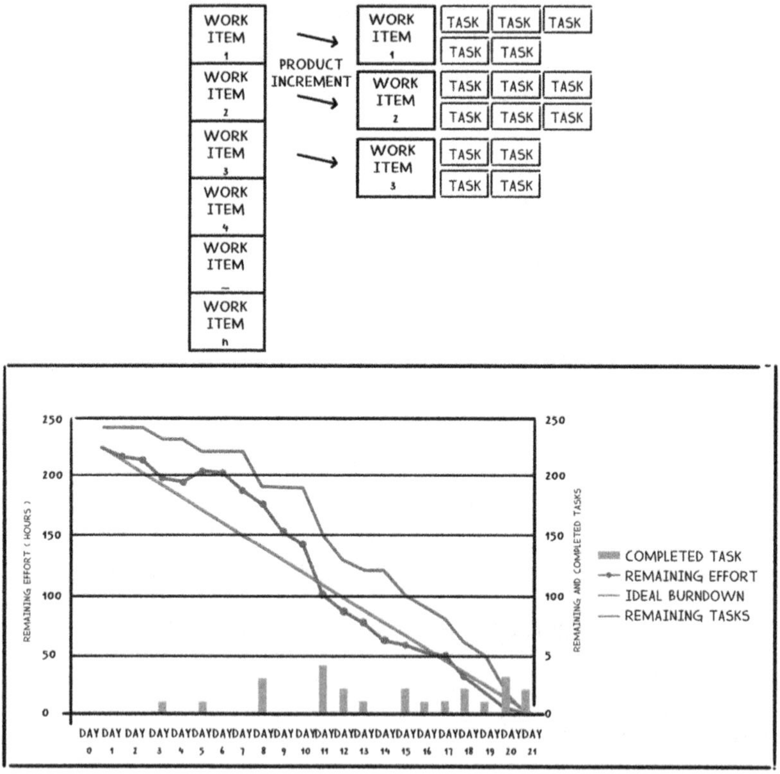

Figure 60 - Product Backlog feeds the Sprint Backlog, which results in Burndown Charts

Simply put, if your product backlog contains 450 user stories representing 2,400 story points and your development team is delivering at a steady velocity of 60 story points for every 2-week sprint, you are able to predict that the rest of the project will take another 40 sprints, which equals 80 weeks, or roughly 1 ½ years.

2,400 / 60 = 40 sprints
40 sprints x 2-week sprint duration = 80 weeks
80 weeks / 52 weeks in a year = 1.54 years

Accordingly, calculating your burn rate becomes a trivial exercise. Simply multiply your labor cost and overhead by the duration. For example, if you know that your 7-member development team costs you $1,529,500 per year ($805,000 for salaries, $724,500 for overhead / fringe benefits, etc.) then you can easily figure out the cost per sprint ($1,529,500 / 26 = $58,827).

You cannot get any better at forecasting and predictability. Agile / Scrum estimation using estimation poker is an empirical, self-correcting, approach to short-term software estimation where user stories are fully defined.

Because it is an empirical approach to project planning, you always know exactly where your project stands at any given time.

T-shirt sizing (affinity estimation) is used for higher level estimates needed for budgeting, staffing, and other longer term planning activities where not all user stories are defined in all detail and a quick assessment of upcoming work is needed.

Chapter 11

Why #NoEstimates gets it #AllWrong

Over the last couple of years, the #NoEstimates movement has gained momentum, and with that momentum misconceptions abound. Not a project goes by where engineers do not start arguing about what, when, how, how much, and – because of the #NoEstimates movement – whether to estimate at all.

This chapter is not a book critique of Vasco Duarte's ***"No Esti-***

mates"[26] – lots of scathing criticism is floating out there on the internet and the blogosphere; but many of the assumptions seem outdated and seem based on questionable research that is 15+ years old.

#NoEstimates is an interesting idea, and in a Lean and Kanban sense of the word one can even see the logic behind it; however, there are fundamental questions whether #NoEstimates can be applied in most projects. Quite the contrary; #NoEstimates is dead on arrival in most corporate environments or in environments that are short on cash.

Most complex projects or software development efforts need to follow a realistic and disciplined approach to creating estimates in Agile / Scrum environments, as explained in some of the chapters of this book:

- Quick & Dirty Guide to Writing Good User Stories
- Agile Estimation How To
- Agile Release Planning
- Managing the Product and Sprint Backlogs
- Delivering Good Product Increments

Estimation is difficult, but good estimation is possible if you follow best practices.

Instead of arguing for not producing good estimates, the industry at large should be focused on enabling better estimates.

[26] https://www.amazon.com/NoEstimates-Measure-Project-Progress-Estimating-ebook/dp/B01FWMSBBK/

Lastly, as mentioned earlier, estimation variance and estimation uncertainty is a fact of life. Nobody, in any industry, can come up with 100% accurate estimates – but you can get pretty close.

THE WORLD WE LIVE IN

Every software product or project is dealing with the following dynamics:

- **Everybody has a boss** – especially in large companies, like companies that are publicly traded and in which you have investments through your 401K portfolio, predictable financial outcomes reign supreme. You would be sorely disappointed if a company projected a $1B profit for this year ending, but ended up with only $500M, resulting in plunging stock prices. Couldn't they estimate their sales better? Contain cost better? What's the matter with them? Software projects in companies need to estimate their effort and ultimately their cost. Everybody else has to; why not software development?
- **Everybody has to live within their budget** – if a home buyer is going to spend $600K on a buying a custom-built home and the general contractor / architect tells the buyer "#NoEstimates, we'll have to see how quickly we generate a viable home for you…," you'd walk away from them. Now if he'd tell the buyer that the cost will end up being between $600K and $650K due to fluctuating resource pricing and weather uncertainties, you'd go for it because that makes sense.

- **Business people and developers must work together daily** throughout the project – understanding the business side better enables software developers to generate better estimates; things will change, and you need to welcome change as a given, but do not assume that changing requirements completely obliterate any estimation effort. People focus on the extremes here, acting as if one minor requirements change invalidates the entire estimate. Yes, sometimes changing requirements do blow the estimate out of the water, completely – but you move on and estimate again based on your new knowledge.
- **Things are not perfect in this imperfect world** – you might not have the best team you want, or you might be distributed across 6 time zones because your company partnered with overseas providers, or you might be stuck on 10-year-old server technology, etc. However, despite all the obstacles, you can still follow best practices, and you can still try your best in estimating the development effort.
- **The reasonableness test** – we expect precise estimates for all kinds of complex endeavors in other industries, but many software professionals are unwilling to try. So, the obvious question is "Why?"

SOFTWARE DEVELOPMENT MYTHS

There are persistent myths regarding software development that enable us to simply take an "estimation is impossible" stand. Some favored ones:

- **Every project is unique and special** – no, not really. You might not have ever done such a project, but most projects have been done before, somewhere. Blogs are full of reference points of similar apps or projects, to the point where people copy each other's code, so the argument that it's unique and special and hence cannot be estimated is bogus. Again, following best practices helps.
- **Software development is an art** – no, software development is at its core an engineering discipline. There might be some disciplines such as UX that are more "artsy" and intuitive, but most of the tasks are engineering tasks, hence the title Software Engineer.
- **"You just don't understand the complexity"** – fair enough, it's unreasonable to ask software developers to estimate something they don't understand, but the solution is not to simply give up but rather to simplify, communicate, interact with the business, have the product owner document good user stories, and come up with an estimate.

Unfortunately, the #NoEstimates movement has confused the subject of generating good, reliable estimates with a catchy hashtag. Too many people now jump to the conclusion that estimates are not necessary any longer, without ever having seriously tried all the best practices that enable good estimates.

Poorly planned, poorly managed, and poorly margined projects will fail, regardless of your estimation techniques – it's really that simple. #NoEstimates will not help in this case either.

Chapter 12
Quick & Dirty Guide to Writing Good User Stories

User stories serve one main purpose: to encourage conversation and discussion about features and functionality documented in the user story. By design, user stories are supposed to be simple and short descriptions of features to be developed, written from the end user's perspective, in the end user's language.

This is important to remember. User stories facilitate conversation and discussion in order to create a common understanding of the desired functionality amongst the team members. User stories are NOT comprehensive specifications.

User stories typically follow a simple template:

As a <user> I want to <action> so that <result / benefit>

Basic understanding of the user story requires all team members to be clear on:

Who? As a <user>
What? I want to <action>
Why? So that <result / benefit>

Every team member is supposed to be able to explain **"who?", "what?", and "why?"** for each user story. It's important to explicitly call this out because unless each team member understands the essence of each story, the team as a whole is lacking the necessary understanding of the problem space. Note that not every team member needs to understand how to implement the user story, but each member should understand the essence.

User stories can be stored in an Application Lifecycle Management tool (such as VersionOne, Jira, Team Foundation Server, etc.), but whenever possible, simple 3" x 5" index cards are preferred.

Especially at the beginning of a project effort, index cards provide an effective way to brainstorm, shuffle things around, and create a high level user story view of what's going on. The tactile nature of holding cards and physically moving them around seems to facilitate early team building and collaboration, something that often is difficult to achieve in front of a screen, no matter what size it is.

This is a similar experience one might have with writing down notes on a piece of paper versus taking notes on a computer. For the majority of people, the process of physically writing down information on a piece of paper greatly improves comprehension and memorization, whereas typing on a keyboard often does not.

Later on, transitioning to an online system makes sense. That, though, is more of a preference, as the author has seen fairly large projects entirely managed on index cards and sticky notes. Obviously, there are many drawbacks to maintaining a stack of cards for months – lack of reliable backup being the biggest one of them.

User stories should follow the INVEST principle. They should be:

- Independent (to the degree possible, user stories should not rely on other user stories to be implemented)
- Negotiable
- Valuable
    - Feature focused, not task oriented
    - Written in the user's language to the extent possible; where domain-specific language is required, such as in genomics or medicine, a subject matter expert will have to be available
- Estimable
- Small
    - Half a sprint or less for one development team member
- Testable

- Needs to have acceptance criteria and not be subjective

The testability of a user story is usually documented in the form of **acceptance criteria** on the reverse side of the index card (or the appropriate place in the ALM tool of your choice) and lists of pass/fail testable conditions that help us verify if the user story is implemented as intended.

All of this sounds simple enough, but frequently teams and product owners really struggle with writing good user stories, and 90% of all problems encountered are based on user stories being too big.

When a user story is too big, it is hard to understand, estimate, implement, and prone to wildly different opinions. So what is a good size for a user story? Basic guidance is that the user stories at the top of the product backlog (the ones that should have sufficient specificity in order to be worked on by a developer) should be sized such that a team member could easily complete two user stories during a sprint.

When a team's user stories are smaller, the team completes stories more frequently. One of the great side effects of smaller user stories is that the team gets more frequent feedback, more frequent successes, and the burndown charts are more granular and look more like a graph instead of a set of stairs.

How does a team take the big stories in its product backlog and split them into smaller stories? There are three basic techniques that we can use to split user stories.

1. Splitting stories with generic words.
2. Splitting stories by acceptance criteria.
3. Splitting stories with conjunctions and connectors.

SPLITTING USER STORIES WITH GENERIC WORDS

This approach simply looks out for words that are open to a wide interpretation (i.e., that are too "generic") or have multiple meanings. For example, if you read a sentence like "We are moving all infrastructure to the cloud," what does that mean? You might have seen or heard sales pitches that sound great, but 2 minutes later you set back and think to yourself "I have no idea what that means!"

What is infrastructure? Servers? Desktops? Laptops? Only database servers? Buildings? Are you getting rid of all your data centers? Some? Move from desktops / laptops to Chromebooks?

What cloud? Amazon's AWS? Microsoft's Windows Azure? IBM's Bluemix? Public? Private? Hybrid?

The point is the generic words lack specificity. Perhaps an example would help explain this:

As a credit card transaction,
We want activities to be logged,
So that the account ledger is up-to-date.

In this story, the word "activities" is pretty generic. We can replace "activities" with more specific words such as: *debits, credits, voided transactions.* We will get these stories.

As a credit card transaction,
We want *debits* to be logged,
So that the account ledger is up-to-date.

AND

As a credit card transaction,
We want *credits* to be logged,
So that the account ledger is up-to-date.

AND

As a credit card transaction,
We want *voided transactions* to be logged,
So that the account ledger is up-to-date.

By providing specificity, you enable to team members to realistically assess what is needed for an implementation, and, more importantly, you avoid ambiguity where individuals can interpret the same user story in different ways.

Generic words are the bane of good user stories!

SPLITTING USER STORIES BY ACCEPTANCE CRITERIA

This method looks at the acceptance criteria listed on the back of a user story card to split user stories into smaller, digestible chunks. What are acceptance criteria? Each user story should have between 6 and 12 acceptance criteria. The product owner works with the team to create, agree upon, and record the acceptance criteria for each user story before the story enters a sprint. Think of this as the "definition of done" at the user story level.

Again, let's look at an example:

As a credit card user,
I want voided transactions to be logged,
So that the account ledger is up-to-date.

Here are some acceptance criteria for this story:

- Transactions support US dollars
- Transactions support Canadian dollars
- Transactions do not support Euros
- Transactions support Visa Cards
- Transactions support MasterCards
- Transactions support American Express cards
- Transactions do not support Diners Club cards
- No transactions above $500

If we examine each one of the acceptance criteria, we can ask "Who wants this?" The answer to this question becomes the user in "As a (type of user)." Next, we ask "Why do they want that?" The answer to this question identifies the value in "so that (some value is created)." The body of the acceptance criteria provides the "I want (something)" part. Here are user stories that could be derived from the acceptance criteria above.

As a Visa credit card user,
I want US dollar voided transactions to be logged,
So that the account ledger is up-to-date.

AND

As a Visa credit card user,
I want Canadian dollar voided transactions to be logged,
So that the account ledger is up-to-date.

AND

As a MasterCard credit card user,
I want US dollar voided transactions to be logged,
So that the account ledger is up-to-date.

AND

As an American Express credit card user,
I want Canadian dollar voided transactions to be logged,
So that the account ledger is up-to-date.

AND

As a Diners Club credit card user,
I want to receive an error message denying the transaction,
So that the user has a clear understanding that Diners Club is not supported.

ETC.

Using acceptance criteria to break up bigger user stories into more digestible ones is simple and quick and helps refine the backlog so that ambiguity is driven out. It also allows the product owner to be more effective in prioritizing work. In our example above he/she might decide to implement Visa card support in Sprint 1, MasterCard support in Sprint 2, but delay American Express card support to Sprint 7 – depending on the business needs and value each provides.

SPLITTING USER STORIES WITH CONJUNCTIONS AND CONNECTORS

This approach simply encourages the team members to review the user stories and look for connector words such as: **and, or, if, when, but, then, as well as,** etc. Commas and semicolons often function as connectors.

It is usually safe to assume that you can split the user story into two user stories by separating the pieces on each side of the connector words, for example:

As a credit card user,
I want US dollar voided transactions to be logged, **as well as** Canadian dollars, **but not** Euros,
So that the account ledger is up-to-date.

You can easily make this into three distinct User Stories:

As a credit card user,
I want US dollar voided transactions to be logged,
So that the account ledger is up-to-date.

AND

As a credit card user,
I want Canadian dollar voided transactions to be logged,
So that the account ledger is up-to-date.

AND

As a credit card user,
I want to receive an error message denying the transaction when trying to void a transaction in Euros,
So that the user has a clear understanding that Euros are not supported.

Of course, each of these in turn can be analyzed based on other criteria (Visa card user, MasterCard user, etc.).

That's pretty much it! Simple. Although there are other techniques to analyze user stories, the three mentioned above will get most teams to write much better, more precise, and smaller user stories.

Because of the more granular specificity, the implementation team will have a better chance of actually implementing the right thing, while providing the product owner with the opportunity to prioritize the backlog in a more optimal way.

Chapter 13
Definition of Ready (DoR) vs. Definition of Done (DoD)

Sometimes referred to as an "Operating Agreement," both the Definition of Done and the Definition of Ready really only serve one purpose, mainly to make it clear when user stories or product increments are ready to be consumed downstream.

THE BIG PICTURE

One of the key artifacts of Agile / Scrum is the idea of backlogs – both at the product level as well as at the sprint level.

The **product backlog** is essentially the project's To-Do list, or requirements repository.

All items that are deemed in scope for the project, regardless of level of detail, are in the product backlog, and they are ordered, not just prioritized – meaning the one on top is more important than the one in 5th position, which is more important than the one in 23rd position. Order is decided by the product owner and usually is driven by business value – in consultation with the development team.

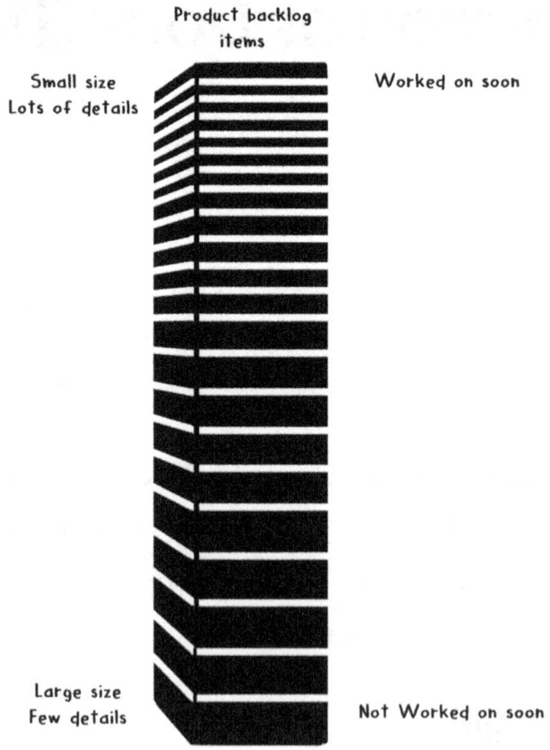

Figure 61 - The Product Backlog

The project's **scope** is written down and documented in the form of user stories, with the expectation that discovery will lead to change. The product backlog is a living document (or repository) and is owned by the product owner.

The product backlog is the source for the sprint backlog. Whereas the product backlog represents the requirements repository for the project, the sprint backlog is the agreed-upon scope for the next upcoming sprint and as such represents the development team's deliverable commitment.

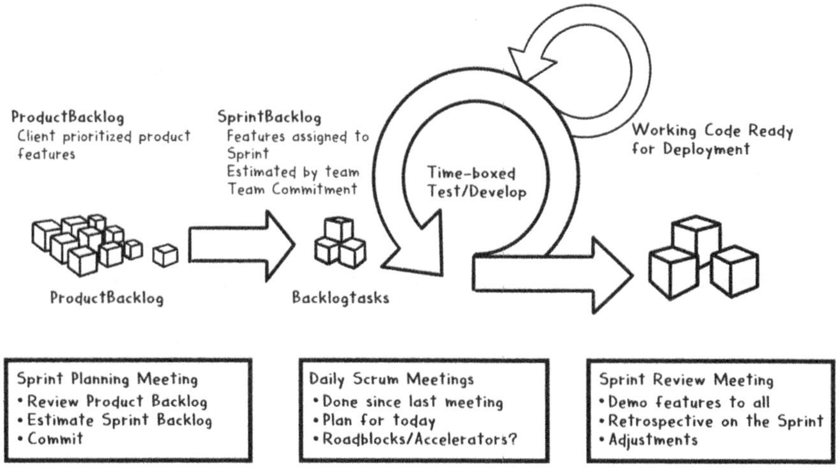

Figure 62 - Agile / Scrum Process

Once agreed upon and committed to, the sprint backlog usually does not change in order to ensure the development team can deliver against their commitment.

The development team works through the sprint backlog top to bottom (most important to least important). If it runs out of work, the team usually looks at the product backlog's top items to pull additional work into the current sprint.

If items are not finished during the current sprint, they revert back to the product backlog to be reconsidered for the next sprint in the upcoming sprint planning meeting.

Newly discovered requirements are added to the product backlog in the form of user stories, to be considered for future sprints.

DEFINITION OF READY VS. DEFINITION OF DONE

A sprint is a time-boxed development cycle that takes high-priority items off the sprint backlog and turns them into a product increment. However, in order to successfully pull items into the current sprint, it is important that the defined user stories are "ready" – pulling unfinished or unrefined user stories into a sprint causes problems during the implementation cycle, as it follows the old principle of "garbage in, garbage out." If developers work off of insufficiently detailed or defined user stories, they are unlikely to produce high quality code.

A "ready" backlog item needs to be *clear, testable, and feasible*:

- A user story is **clear** if all Scrum team members have a shared understanding of what it means. Collaboratively writing user stories, and adding acceptance criteria to the high-priority ones, facilitates clarity.
- An item is **testable** if there is an effective way to determine if

the functionality works as expected. Acceptance criteria ensure that each story can be tested.

- A user story is **feasible** if it can be completed in one sprint, according to the Definition of Done. If this is not achievable, it needs to be broken down further.

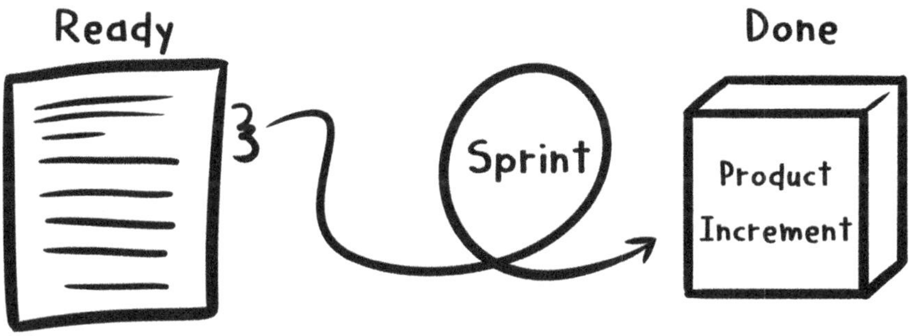

Simply stated, the DoR defines the criteria that a specific user story has to meet before being considered for estimation or inclusion into a sprint.

Whereas a DoR is focused on user story level characteristics, the DoD is focused on the sprint or release level. Essentially, a DoD represents the acceptance criteria for a sprint or release. It spells out what the development team has to cover in order for the product increment to be considered "done."

The DoD is an agreement between the development team and the product owner on what needs to be completed for each user story. It is often standardized across the company in order to guarantee consistent delivery of quality.

Things commonly addressed in the DoD are:

- **Operating environments** and at what level of **integration** are user stories expected to work (what specific version of Linux, what specific version of Android, iOS, or browser)?
- What level of **documentation** is required (automatically generated Javadoc vs. fully edited end user documentation)?
- What are the **quality** expectations (basic functionality works for demo purposes vs. fully designed and bulletproof app)?
- What are the **security** expectations (no security implemented vs. security vetted at all levels, from code reviews, code scans, up through network security configuration)?
- **Scalability** expectations (scalable for demo purposes up to 10 concurrent users vs. scalable to 100,000 concurrent users)?

Essentially the DoD is the agreed-upon acceptance criteria that the product owner will use to accept the product increment at the end of the sprint.

Please note that the DoD may be different for sprints vs. releases. Intermediate sprints might have a less stringent DoD than the final couple of sprints before you are planning to release to market.

Sample Definition of Ready (DoR)

- User story is clear.
- User story is testable.
- User story is feasible.

- User story is defined.
- User story acceptance criteria are defined.
- User story dependencies are identified.
- User story sized by development team.
- Scrum team accepts user experience artifacts.
- Performance criteria identified, where appropriate.
- Scalability criteria identified, where appropriate.
- Security criteria identified, where appropriate.
- Person who will accept the user story is identified.
- Team has a good idea what it will mean to demo the user story.

Sample Definition of Done (DoD)

- Code produced (all 'to do' items in code completed).
- Code commented, checked in, and run against current version in source control.
- Peer reviewed (or produced with pair programming) and meeting development standards.
- Builds without errors.
- Unit tests written and passing.
- Deployed to system test environment and passed system tests.
- Passed UAT (User Acceptance Testing) and signed off as meeting requirements
- Any build / deployment / configuration changes are implemented / documented / communicated.
- Relevant documentation / diagrams produced and / or updat-

ed.

- Remaining hours for task set to zero and task closed.

Other uses of the Definition of Ready and Definition of Done

The original intent of DoR and DoD was to create brief documents representing internal operating agreements among the project's stakeholders, the product owner, and the development team.

However, with more development work being outsourced or subcontracted, the DoR and DoD are now used more and more in contract agreements and statements of work, spelling out exact expectations as to what needs to be done.

These are useful tools for negotiating project scope, as they define expectations and hold both parties accountable. The DoR helps the customer in producing well-written user stories that are ready to be consumed by the development team. The DoD helps the implementation partner in producing working product increments according to all project requirements, not just the specific user story functionality.

Chapter 14
Effective Daily Scrum Meetings (aka Daily Standups)

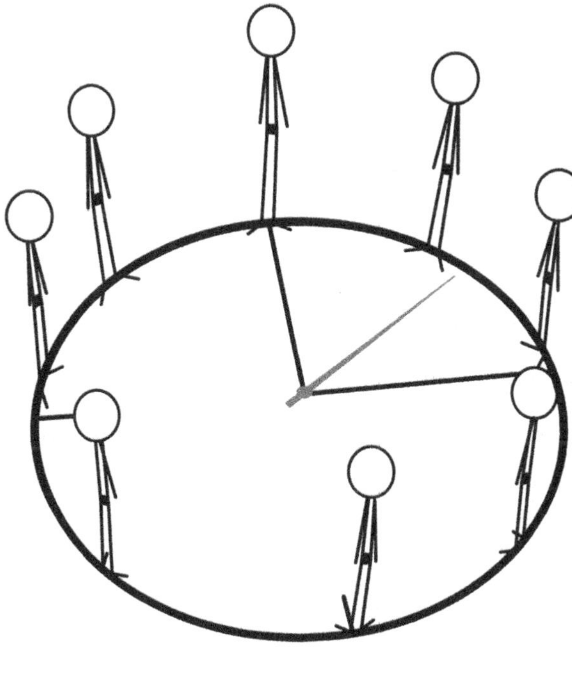

Figure 63 - Daily Scrum Meeting

SIMPLE RULES

Daily Scrum Meetings (aka Daily Standups) have very simple rules that can be easily followed:

1. The meeting is time-boxed at 15 minutes! No exceptions!
2. The scrum master runs the meeting and acts as the meeting facilitator.
3. It is recommended to have participants stand in order to keep it short.
4. Development team members focus in on only three key areas of communication:
5. "This is what I did yesterday…,"
6. "This is what I can do today…," and
7. "These are the things impeding me… – I need help!"
8. The scrum master talks to the status of the impediments log – especially any issues that have been brought up as impediments by team members before and have immediate impact on productivity.
9. The meeting is focused on coordination; problem solving should be done outside of this meeting.
10. Only development team members, the scrum master, and the product owner can talk – all others can observe or listen in.

RELEASE	SPRINT	CATEGORY	IMPEDIMENT	STATUS	RAISED ON	RAISED BY	NOTES

Figure 64 - Simple Running Improvements Log Used in Daily Scrum Standup

COMMON CHALLENGES AND HOW TO DEAL WITH THEM

Stakeholder Discussions

As mentioned, only development team members, the scrum master, and the product owner can talk – but what do you do if the VP of Engineering or CEO decides to show up?

This is where a good scrum master is worth his money. Good scrum masters know how to effectively navigate a situation like this, and make sure everybody knows the rules. Ideally the scrum master talked to all stakeholders beforehand to set expectations.

It is important to make sure that stakeholders attending the daily scrum meeting understand that they can listen and observe, but they cannot participate in the discussion. This is critical for four main reasons:

1. To optimally use the available 15-minute time window.
2. To keep momentum.
3. To keep focus.
4. To maintain team "flow" – executive level stakeholders actively jumping into the coordination discussion is more disruptive than helpful.

Team Drifts into Problem Solving

Development teams sometimes have a tendency to drift from coordination and status discussions to actually trying to discuss potential problem solutions.

Again, a good scrum master needs to step in here and gently but firmly drive the point across that the daily scrum meeting is for coordination purposes only, not a working session to discover solutions to problems.

Simply stating something like "*Ok, Paul, why don't you take that offline with Peter and Mary after the meeting?*" usually does the trick.

On-Time Attendance

Every team has their own happy place when it comes to meeting times. Some teams love early morning meetings to get the day started at 7:00 AM. Others like a 9:30 AM or 10:00 AM time slot.

Whatever works for the team should be the time you pick. It can be mornings, afternoons, or even around lunch, as long as it is daily.

However, once a time is decided upon, all team members are expected to show up on time. It makes little sense to have a 15-minute meeting if one-third of the participants are 10 minutes late. That defeats the purpose.

The scrum master has to figure out with the development team and the product owner what the best time is and then work with the team to make sure there is the appropriate discipline around attending on time.

Team distribution across several time zones is one deciding factor here as well – see below.

Filibusters

Considering the normal Scrum team size of 7 (+/- 2), you are looking at roughly 2 minutes allotted time that each team member has to share yesterday's accomplishments, today's plan, and any impediments. Not a lot of time.

Beware of long-winded status updates from team members. Keep it short and sweet and to the point. No filibusters!

Team Distribution across Time Zones

The following picture shows the typical Scrum meetings across two time zones, US and India, assuming 2-week sprints.

Before the sprint starts you need to have a sprint planning meeting. Once the sprint is underway, you need to conduct the daily scrum meetings. You finish off the sprint by doing your sprint review meeting and your sprint retrospective meeting.

This is a typical setup for a team that is distributed. This kind of setup is far from ideal, but it is workable as long as you have the right communication equipment for video- and teleconferencing.

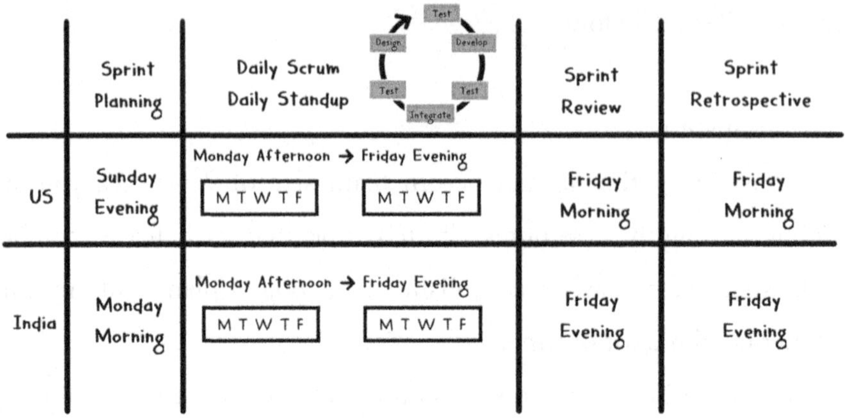

Figure 65 - Agile / Scrum Events - How to deal with time zone distribution

Facilities

Team meeting rooms, whiteboards, video- and teleconferencing equipment all are necessary for a scrum team to operate effectively.

Companies dedicated to Agile can be productive with 100% remote setups because online collaboration tools such as WebEx, Lync / Teams, HipChat, and Skype allow face-to-face interaction. Again, 100% remote situations are not ideal, but they are becoming more and more workable with the right technology tools.

CHAPTER 15
EFFECTIVE SPRINT REVIEW MEETINGS

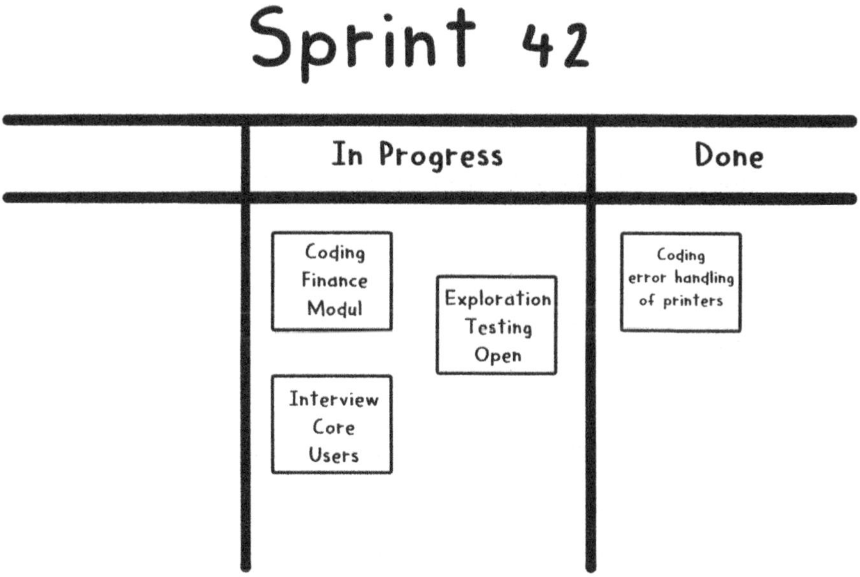

Figure 66 - Sprint Review Discussion

In Scrum, sprint review meetings are review meetings in which the product owner presents the progress that has been made in the current sprint.

These meetings occur at the end of every sprint, allowing the product owner to formally accept the product increment – or not – based on the agreed-upon Definition of Done (DoD).

Sprint review meetings represent an opportunity for stakeholders or other interested parties to observe the "real thing."

Inputs

- Definition of Done (DoD)
- Sprint goal, as expressed in a ready set of distinct user stories agreed upon in the sprint planning meeting

Outputs

- Shippable / deployable product increment

SIMPLE RULES

- The product owner presents:
    - The sprint goal, as agreed upon during the sprint planning meeting.
    - What the development team accomplished during the sprint, based on the agreed-upon Definition of Done (DoD).
- The development team demonstrates the working product increment:
    - Only user stories that the product owner has approved

    are considered done.
    - Within the context of a single sprint, a product increment is a work product that has been done and is deemed "shippable" or "deployable" – meaning it is developed, tested, integrated, and documented – as agreed upon in the DoD.
- The sprint review meeting is time-boxed to 1 hour per week of sprint duration.
- This is an informal meeting – it is not a corporate song and dance! It requires only minimal preparation time and demonstrates real-life, functionality (no PowerPoint slides, mockups, or rigged demos). This informality is an important point, as you do not want development team members to spend hours or days preparing a corporate-level presentation.
- The main purpose of this meeting is to show the reality and current state of the development effort at the end of the sprint.
- A product increment may or may not be released externally, depending on the release plan.

COMMON CHALLENGES AND HOW TO DEAL WITH THEM

The Product Owner is surprised about the State of the Product Increment

Product owners are supposed to be dedicated to the development team and interact with the team on a daily basis. Participation in daily scrum meetings is mandatory for the product owner, and those daily meetings usually expose any potential surprises. If that is the case, the product owner knows exactly where the sprint stands in terms of the product increment functionality.

However, the reality is that many product owners have "day jobs," and as such, often do not have the bandwidth to participate in all Agile / Scrum events and meetings. Consequently, they are sometimes "out of the loop" on the latest coding challenges. This can lead to the product owner being surprised as to what functionality was finished or not while participating in the sprint review meeting.

In order to remedy this, make sure that:

1. The product owner is dedicated full-time to the effort.
2. The product owner attends all necessary Agile / Scrum coordination meetings (sprint planning, daily scrum / standup, sprint review, and sprint retrospective).
3. Functionality is developed throughout the sprint, avoiding a big wave of code to be finalized a day before the sprint review – and produce potential slips.
4. The scrum board is updated with the latest information; it is a good practice to start off the sprint review meeting with a quick review of the scrum board – if functionality is not working, it is not considered done.

There Is Nothing to Demo

Depending on the development effort, teams sometimes struggle with product increments that cannot demonstrate functionality visually.

Good examples are the development of a backend, server-side, API or a predictive analytics model that produces a cryptic output. How do you show your progress?

There are several things you can do to visualize progress:

Utilize tools / visualization aids that you already use in your development effort – for example, if you are developing an API, you can use mockups and unit test frameworks to demonstrate functionality.

Make cryptic results and complex input data more easily understandable – for example, if you have a complex scoring algorithm that produces a result based on a large, varying, data set, use some kind of visualization, like Excel Pivot Charts, to summarize the data while demonstrating the functionality.

Talk people through the less visually exciting parts of the functionality – not ideal, but sometimes the best you can do is talk the audience through the functionality when visually it does not dazzle.

Utilize as many existing tools and validation processes as you can in order to avoid adding additional "visualization tasks" to your already busy delivery schedule. Many times developers / testers might have tools that can be used to show the less visible functionality.

The Product Increment Crashes and Burns

This should never happen, but sometimes it does! If the development team uses continuous integration and automated unit testing, the basic stability of the system should be guaranteed.

If, for whatever reason, a late-breaking change introduces instability, it is better to cancel or reschedule the sprint review meeting.

The reason for the latter is simple: a) to not waste other people's time, and b) to avoid creating a bad impression.

It is better to say "It's not working yet, we have to reschedule" than to sit in a meeting and say "Oops, I don't know why that's not working."

As a matter of process, the development team and the product owner should test drive the product increment before showing it during the sprint review meeting.

A general guideline to follow is to finish up all development, testing, integration, and documentation activities the day before the scheduled sprint review meeting, in order to give the team enough time to make sure everything is in good order.

The Sprint Review Turns into a Political Showdown

Beware of stakeholders and other influencers in your organization who may use the sprint review meeting as a battleground for a political showdown. If your organization has not fully transitioned to an Agile / Scrum process, stakeholders and influencers might view the sprint review meeting as another kind of project review or phase gate review meeting more common in waterfall / PMBOK aligned processes.

This is where a good scrum master and product owner are worth their money.

Good scrum masters and product owners know how to effectively navigate a situation like this, and to make sure everybody knows the rules. Ideally, both the scrum master and the product owner have talked to all stakeholders and influencers beforehand to set expectations.

It is important to make sure that stakeholders attending the sprint review meeting understand that they can listen and observe, but they cannot abuse the meeting to carry out political hit jobs on the project. The development team should not get exposed to whatever political / strategic / budgetary disagreements may be brewing behind the scenes, as it causes the technical staff to get disillusioned and question the leadership.

The key point to remember here is that the sprint review meeting is intended to provide a clear and unvarnished view of the "as is" state of the current product increment. It is focused on the sprint deliverable. In the spirit of complete transparency and open communication, the sprint review meeting might expose shortcomings or problems – that is its intent.

The focus of the sprint review meeting is just that – to review the sprint and the associated deliverables, the product increment. The meeting is not intended as a project review meeting or a phase gate checkpoint meeting common in waterfall / PMBOK processes.

CHAPTER 16
EFFECTIVE SPRINT RETROSPECTIVES

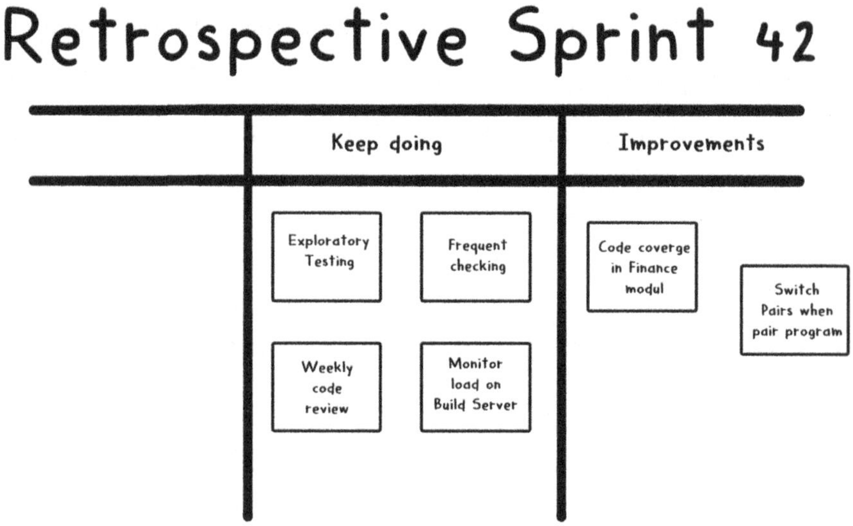

Figure 67 - Sprint Retrospective Discussion

Sprint retrospective meetings have a simple intent: To discover, discuss, and take note of tangible process improvement opportunities.

A sprint retrospective is essentially a process improvement meeting at the end of each sprint where the Agile / Scrum team discusses what went well, what should change, and how to implement any changes.

SIMPLE RULES

1. The scrum master acts as the meeting facilitator.
2. The sprint retrospective scope is focused on the most recent sprint.
3. Meeting participants include:
    - Scrum master
    - Product owner
    - Development team
    - If agreed to by the team, it is possible to have other participants, such as customers or other stakeholders, but that is entirely up to the team to decide.
4. During the sprint retrospective, the team focuses on the essential three questions:
    - What went well?
    - What would we like to change?
    - How can we implement that change?
5. The key to making a sprint retrospective useful is to keep it "action focused" – the team needs to focus in on what they can change to make their own process better.
6. The meeting time is fixed: 45 minutes for each week of sprint duration.

MANAGING PROCESS IMPROVEMENTS OVER TIME

As mentioned, the sprint retrospectives should be **action focused** – the intent is to rapidly improve upon the process in order to enable better productivity.

But, the reality is that some improvement suggestions might be quickly addressed, some might take some time, and others might be outside of the team's area of decision power.

It is suggested to keep a running log of improvement suggestions that will help you keep track of what improvement ideas came up, what category they fall into, and what sprint / release they first came up in, etc.

This helps keep the team accountable and ensures you remember all the process improvements you implemented over time.

RELEASE	SPRINT	CATEGORY	IMPEDIMENT	STATUS	RAISED ON	RAISED BY	NOTES

Table 8 - Simple Running Improvements Log Used in Sprint Retrospectives

The time frames that some improvement suggestions fall into are as follows:

Within the Next Sprint

An improvement suggestion that can be accomplished over the course of the next sprint, for example, refactoring of a specific web service interface. Quick hits, quick successes.

Over Several Sprints

Improvements that might take two or more sprints to implement. Re-architecting subsystems, or a comprehensive UI update, tasks that cannot be finished in one sprint, fall into this category.

Over the Next Release

Infrastructure refresh efforts, such as switching continuous integration (CI) tools / environments, might fall into this category. Something that you coordinate by release so you do not interrupt the delivery flow of your sprints.

Never

Yes, some people do not like the idea that some improvement suggestions will never be addressed, but we all know situations like this. Say, for example, you are working on a huge legacy system written in Java/Linux, totaling 3 million lines of code. And the improvement suggestion is to rewrite the entire system in C#/Windows.

This is an improvement suggestion that falls into the "Never" category, not because it might never happen, but because it is a decision in terms of impact and scope that cannot be decided upon by the team.

COMMON CHALLENGES AND HOW TO DEAL WITH THEM

Boiling the Ocean

It is incumbent on the scrum master to keep the team on track during sprint retrospectives. Sometimes teams have a tendency to focus on problems that fall into the "boiling the ocean" category.

Those are sometimes great discussion topics (especially for a long night over a couple of beers), but they almost always fall outside of the scope and sphere of influence of the team.

In order to make sure that the sprint retrospective is useful and produces good improvement suggestions, the team should avoid these kinds of discussion topics and focus on two parameters to guide the discussion:

1. Is the improvement suggestion relevant to the scope (last sprint / current release)?
2. Is the improvement suggestion achievable by the team?

Non-Participation

Several years ago a team member refused to participate in a sprint retrospective. Asked why, he said "Sprint retrospectives are supposed to be like 'lessons learned' meetings… but we never learn any lessons; we still got the same issues we had 2 years ago, so why waste time?"

His point was completely valid. To make sprint retrospectives work, you need to focus on what is relevant (last sprint / current release) and what your team can actually implement. Keeping track in a log as to what improvement suggestions were brought up, which ones got implemented, etc. keeps the team accountable but also shows the progress.

Blame Game

Improvement suggestions should never smell like blame is being assigned. It is important to have open team communication, and to honestly assess what can be improved upon by the team. Be careful not to assign blame but rather focus on potential solutions.

Everything is Great / Bad Attitude

Beware of both irrational exuberance and downtrodden doomsday forecasts. Both extremes are usually unrealistic. You can always improve on the process, and things are never as bad as some folks might make them seem.

No Improvement from Last Retrospective

Once again, track what you want to improve on, and if it is within the scope of one sprint, put it into the backlog for the next sprint.

As a general guidance, try to reserve up to 20% of capacity in early sprints for process improvements. In later sprints that might go down to 10%. Point here is that you have to account for this work. It is not going to happen by itself, or in addition to the work of an already fully utilized team.

Process improvement activities are part of the sprint activities, not in addition to it.

CHAPTER 17
AGILE TEST AUTOMATION

The term "automation" is everywhere nowadays, and within the Agile / Scrum community "DevOps" has become a summary buzzword for all kinds of automation activities along the delivery pipeline of software development. Some of the DevOps activities are summarized in the chapter "***Agile Adoption Pitfalls – PART 7, Ignoring DevOps.***"

Please note that **DevOps automation** and **test automation** should not be conflated – they are different activities with a set of different tools, and they often are relevant to different parts of the software development life cycle. This chapter lays out a high level summary of test automation in an Agile environment.

Disclaimer

This chapter is focused on the author's recent experience with modern browser-based and mobile platform (Android, iOS) test automation questions. Legacy development techniques – Windows fat client tests, mainframe green screen tests, and other legacy UI technologies such as Adobe Flash or Microsoft Silverlight – are out of scope.

The focus is mostly on organization and process – how should you and your team go about test automation in an Agile / Scrum environment? It is less focused on the technical nuts and bolts of implementing reliable, scalable, test automation or specific implementation challenges, like how to implement robust automated testing with bulletproof exception handling. So you will not find any code snippets in here.

OPEN ENDED CHALLENGES: THE NEED FOR RISK MANAGEMENT AND PRIORITIZATION

The bad thing about software testing is that there is a real possibility of never, ever, getting to the finish line. Somebody asked once to show a deterministic model of how many tests there would be for a given application. He wanted to know, based on a formula, how many tests would have to be created. Basically, he was looking for a way to determine that application X needed 1,937,297 test cases. There might be models like that out there, but the author has never seen them used in the real world.

In a deadline-driven delivery environment, the *essence of testing is to execute as many tests as possible in as little time as feasible.* That is why test automation is such a significant enabler of Agile / Scrum and DevOps.

"*As many as possible*" is bounded by how fast your test automation can run on servers, execute tests, and log results.

"*As little time as feasible*" is based on your project needs. Instantaneously running automated unit tests every time somebody checks in code is now commonplace. Running extensive automated load tests over days and weeks is also standard practice. As fast as possible, considering your environment, budget, and business needs.

A term that has been used in this context is "turning the test bucket," meaning: how long will it take to run all your tests, on all your supported platforms, at least once? The main reason for this is **confidence**. If we turn the test bucket at least once, we have reasonable confidence that things will work.

There always seem to be more things to test, more things to worry about. Despite the best Definition of Done, if a problem shows up after the fact, then the first question that is thrown out is "Why didn't our tests catch that?"

Consequently, in order to survive, you need to understand and vigorously use **risk management** and prioritize your work using **Pareto analysis** (aka the 80/20 principle).

Risk management is concerned with assessing, avoiding, and controlling risk:

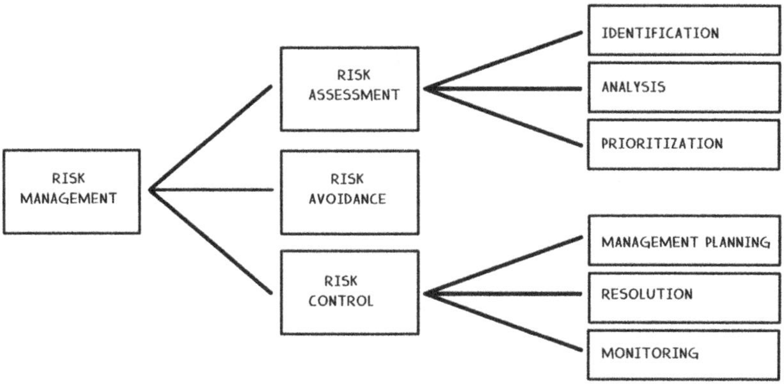

Figure 68 - Risk Management - Assess, Avoid, Control

There is nothing more to it, really. You manage risks by assessing them and then controlling them. The best way of dealing with risk is to avoid it altogether. For example, if you do not want to deal with the risk of earthquakes, put your data center in a place that has very low risk of experiencing earthquakes – that is avoidance.

In order to effectively manage risk, root cause analysis helps identify the key **risk areas** according to the **Pareto principle** (80/20 principle):

RISK	RESOLUTION APPROACH
80% OF SOFTWARE COST IS CONSUMED BY 20% OF THE COMPONENTS	STRIVE TO UNDERSTAND THE DRIVING REQUIREMENTS COMPLETELY BEFORE COMMITTING RESOURCES TO FULL-SCALE DEVELOPMENT
80% OF THE ENGINEERING IS CONSUMED BY 20% OF THE REQUIREMENTS	ELABORATE THE COST-CRITICAL COMPONENTS FIRST
80% OF THE ERRORS ARE CAUSED BY 20% OF THE COMPONENTS	ELABORATE THE RELIABILITY-CRITICAL COMPONENTS FIRST
80% OF THE SOFTWARE SCRAP IS CAUSED BY 20% OF THE CHANGES	ELABORATE THE CHANGE-CRITICAL COMPONENTS FIRST
80% OF THE RESOURCE CONSUMPTIONS (DISK, MEMORY) IS CAUSED BY 20% OF THE COMPONENTS	ELABORATE THE PERFORMANCE-CRITICAL COMPONENTS FIRST
80% OF THE PROGRESS IS MADE BY 20% OF THE PEOPLE	HIRE DELIVERY ORIENTED SOFTWARE PROFESSIONALS THAT CAN FIND A BALANCE BETWEEN DOCUMENTATION PROCESS, AND PRACTICAL ENGINERING APPROACHES

Table 9 - Pareto Analysis Example 1

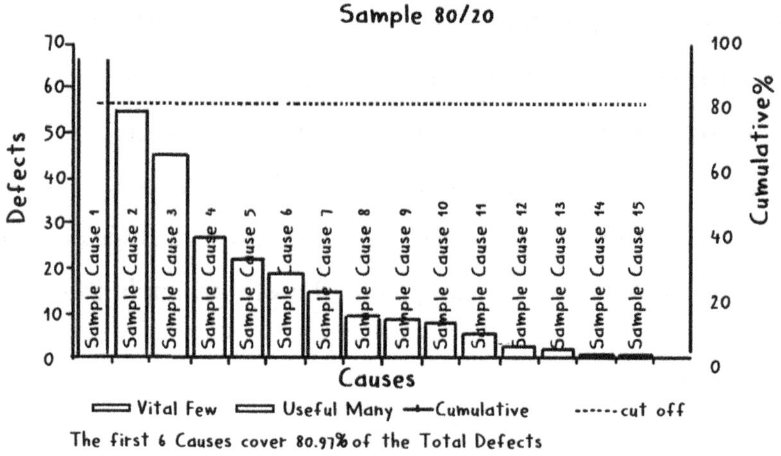

Cumulative Percentage Cutoff 80%

#	Causes	Defects	Cumulative%
1	Sample Cause 1	66	22.8
2	Sample Cause 2	65	41.9
3	Sample Cause 3	45	57.4
4	Sample Cause 4	27	66.8
5	Sample Cause 5	22	74.4
6	Sample Cause 6	19	81.0
7	Sample Cause 7	15	86.2
8	Sample Cause 8	10	89.6
9	Sample Cause 9	9	92.7
10	Sample Cause 10	8	95.5
11	Sample Cause 11	6	97.6
12	Sample Cause 12	3	98.6
13	Sample Cause 13	2	99.3
14	Sample Cause 14	1	99.7
15	Sample Cause 15	1	100

Table 10 – Pareto Analysis Example 2

Effective risk management and Pareto analysis are at the core of efficient resource management and will guide you in selecting the most viable test automation targets.

Unless you prioritize, you will run the risk of having a bottomless pit of tests to run, and to automate, which essentially means you would need unlimited resource – and that usually does not happen on most development projects. Also consider that even if you had unlimited resources, you still would need to prioritize what tests are more important than others in order to get the best use out of your automation.

WILL WE STILL NEED MANUAL TESTING?

Yes. There are several reasons for this:

1. You cannot automate what does not work yet. System stabilization will take some time, and trying to automate while the system is still unstable is like building a house on quicksand.

2. You will continue to need manual testing in order to simulate the end user experience; test automation does not give you critical feedback on how user friendly your software is.
3. Exploratory testing is really the basis for test case analysis, the manual test execution and documented test cases often serve as input for test automation.
4. There are complex application areas that you might never be able to automate, so somebody will have to test it manually.

Many folks believe that automation replaces manual testing. It does not. It supplements it.

For related information and how to calculate the return on investment for test automation efforts, see the ROI Calculations section below.

IMPLEMENTING AUTOMATED TESTS = CODING APPLICATION FUNCTIONALITY

Automated testing (at whatever level) requires software engineering skills. One would argue that mid- to senior-level skills are required for test engineers implementing tests on a full-time basis. You also will need senior- to principal-level skills to architect large-scale automation. Test automation does not happen as a byline in a project. The companies that have been successful implementing extensive automated test suites hire full-blown software engineers with BS or MS degrees, and 5+ years of hands-on coding experience.

As with all complex things, there is a wide spectrum here. From simple automated scripts for a straightforward WordPress site, over a medium sized ERP implementation, to a full-blown e-commerce site that transacts billions of dollars worldwide, you will need a different organization and structure for your test automation effort. This might range anywhere from one software engineer working on a Scrum team coding away simple tests, to an entire team of software test engineers who work closely with DevOps and System Architects. For a perspective on how complex things can get, check out "How Google Tests Software," by Whittaker, Arbon, and Carollo.[27]

In order to successfully implement automation, you need to allocate time for test automation in each sprint. Usually there is not enough time within a sprint to automate the tests for that sprint, so follow the "**Sprint+1 Law**" – this also allows functionality to stabilize and settle after it has gone through some manual / exploratory test cycles.

[27] https://www.amazon.com/Google-Tests-Software-James-Whittaker-ebook/dp/B007MQLMF2/

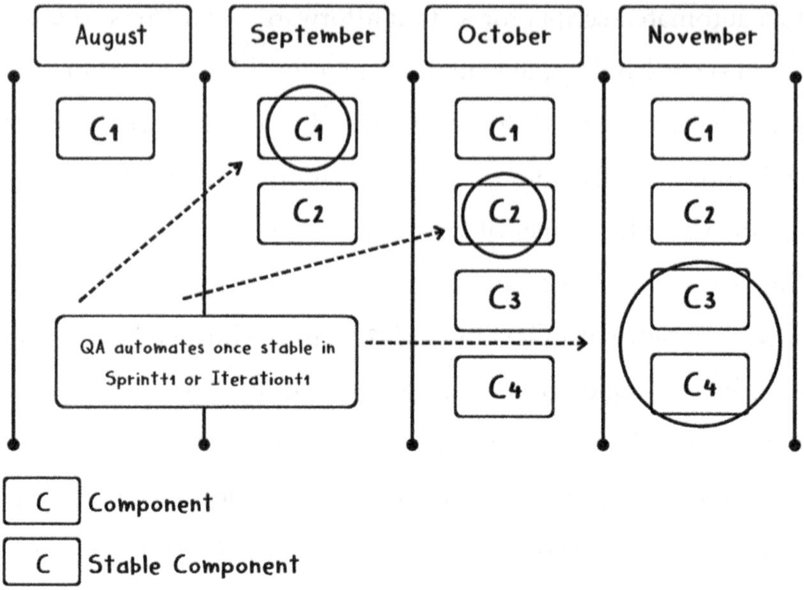

Figure 69 - Sprint + 1 / Iteration + 1 Approach to Test Automation

Do not try to automate components that have not stabilized yet. It is a waste of time. At a minimum, all unit tests need to have passed successfully, and if components are consumed by other software or provide services, the first level of integration should have been done as well. Remember, test automation is intended to speed up the test cycle, not act as the first integration guinea pig.

Do not have automated test creation lag by more than one sprint or iteration ("out of sight, out of mind").

Deal with test automation just as you would with other coding tasks. Express the work in user stories, estimate the work in story points, and track progress using velocity as part of the regular Agile / Scrum process.

If the team is not able to complete its target automation, the gap in remaining automation work is treated just as a functionality coding gap would be, that is, either

a) the sprint is extended to complete the work, or

b) the outstanding tasks are moved to the backlog for allocation to future sprints.

WHAT TO AUTOMATE

Now that you know how to effectively manage the risks and prioritize based on Pareto analysis, you need to start worrying about what to automate.

The following picture provides a pretty good idea of what you should be zeroing in on within the context of a modern internet app.

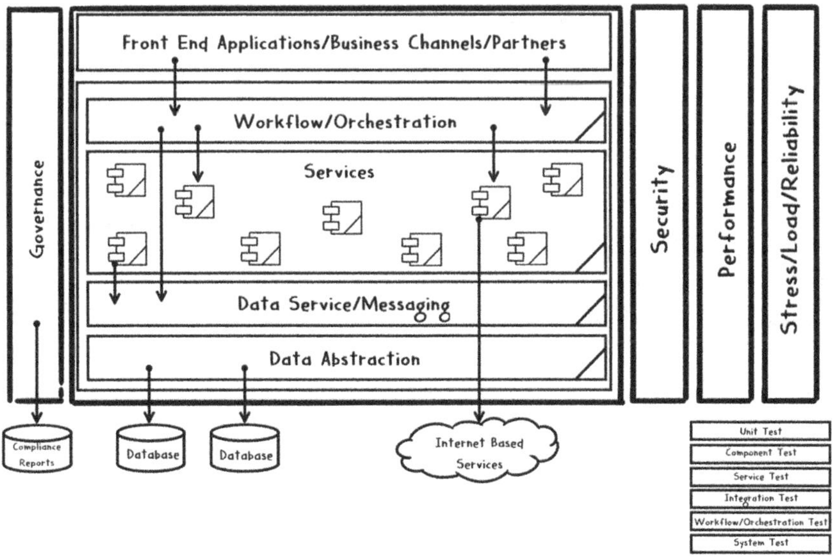

Figure 70 - System Architecture Drives the Categories of Test

Not all areas need to be addressed in all apps; it is entirely dependent on your specific development effort. For example, if you do not have any governance requirements, then you do not need to test for governance compliance – on the other hand, if you need to be HIPAA compliant, you need to worry about that a lot.

The idea is that various test activities provide progressive stabilization of the software under test.

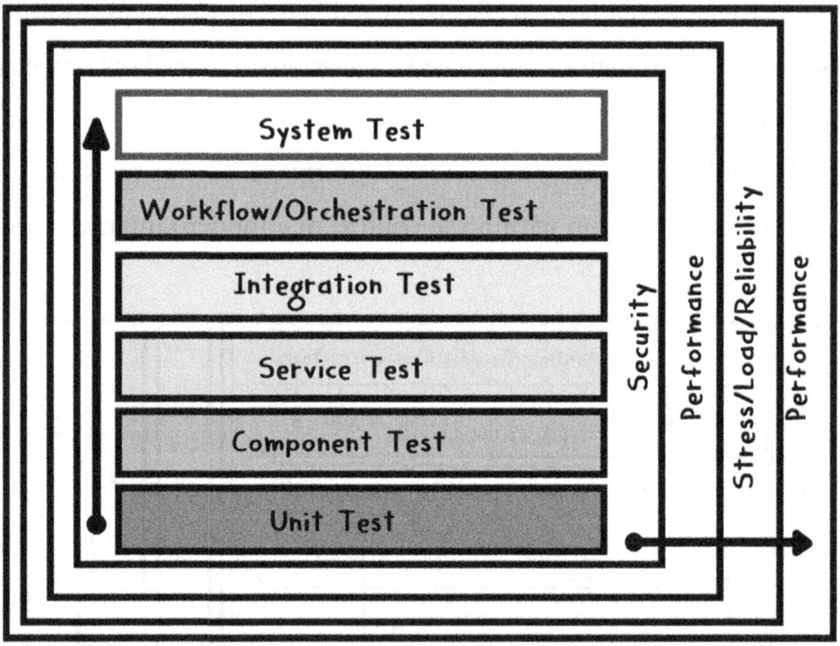

Figure 71 - Different Categories of Test

What is crucial to point out here is that all the automation depends on progressively stabilizing the system, and the basis for this is automated unit tests. Automated unit tests are the foundation upon which other automated tests are written. In the picture above, you automate from the bottom up, and from left to right.

Too many times test automation efforts fail because they tried to automate at the system test level without having the corresponding unit tests automated in the first place. That is usually a sign of a big bang test effort / automation effort at the end of the cycle, and usually spells doom.

Unit tests need to be checked-in along with the code. The idea of checking-in unit tests into the source control system is to couple code changes to an automated test program (usually an automated regression test), allowing for fully automated regression test runs. During the last 10 years, **continuous integration** and commercially available continuous integration servers (Hudson, Continuum, Cruise Control) have become commonplace. And if you are seriously considering continuous integration, then there is no other way to do this, as unit tests are supposed to run with every compilation / build cycle. Each code check-in should be accompanied by a corresponding unit test.

Out of all test automation targets, unit tests provide the biggest bang for the buck!

Different technologies provide different unit testing frameworks that enable easy integration into the build process (native unit test frameworks support MS Visual Studio / TFS, Eclipse, and all kinds of other environments, like Junit for Java, CppUnit for C++, nUnit for various .Net languages, PyUnit for Python, etc.). There is really no longer any reason not to have automated unit tests.

COMBINATORIAL EXPLOSIONS – WHY TO AUTOMATE

Besides the common reasons that drive DevOps automation – like optimizing the software delivery pipeline, creating efficiencies, improved software quality, and development productivity – there is another major reason why we want to automate tests: **the combinatorial explosion / proliferation of platforms**.

Today's products often require multi-platform testing, either to cover various OS versions and variations (Windows, Mac, Linux, UNIX), browser combinations (IE, Firefox, Chrome, Safari), mobile platforms (Android, iOS) or general desktop/mobile combinations. The net result is that testing on multiple platforms has become a necessity for most products, even for internal IT solutions.

Multi-platform testing has become an increasingly large burden on test organizations that have to deal with the combinatorial explosion of systems under test. In Agile / Scrum teams working against rapid delivery cycles, this easily can become the most burdensome effort.

Following is a sample platform browser / device support matrix for an imaginary web site that also provides native apps on Android and iOS:

	Native	Internet Explorer	Firefox	Chrome	Spartan	Safari
Windows 7		X				
Windows 8		X				
Windows 10					X	X
Mac OS						X
Android - LG Revolution	X			X		
Android - LG Optimus V	X			X		
Android - LG Nexus 6	X			X		
Android - Motorola Atrix	X			X		
Android - Motorola Droid Pro	X			X		
Android - Motorola Droid X	X			X		
Android - Samsung Galaxy	X			X		
Android - Samsung Galaxy	X			X		
iOS - 9 on iPhone 6	X					X
iOS - 9 on iPhone 6s	X					X
iOS - 9 on iPhone 7	X					X
iOS - 9 on iPhone 7s	X					X
iOS - 10 on iPhone 6	X					X
iOS - 10 on iPhone 6s	X					X
iOS - 10 on iPhone 7	X					X
iOS - 10 on iPhone 7	X					X

Table 11 - Platform Proliferation and Combinatorial Explosion

That is 37 platform combinations, which is a considerable risk and burden for the team developing the system. If implemented correctly, automation can help with this. With no automation, you are faced with testers trying to cover things manually as best they can. Even sizeable QA groups have challenges covering a large number of platforms like this, so without automation you are essentially rolling the dice and accepting some level of defect leakage to the market.

WHAT TOOL TO AUTOMATE WITH

For a list of current automation tools / frameworks, see Wikipedia.[28] The list is ever changing.

At a minimum, depending on your specific application or system

[28] https://en.wikipedia.org/wiki/Test_automation

space, you need to think about:

- The **coding level** – a tool such as nUnit[29] for unit test automation (successfully used on many programming languages such as Java, C#, C++, Python).
- The **API or web services level** – tools such as SOAPUI[30] for Web Services automation.
- The **UI level** – tools such as Selenium[31] for functional browser automation.
- The **app / mobile level** – tools such as Appium[32] for mobile test automation.

With the same set of tools, you can experience great productivity gains or complete and spectacular failures! Hence, the repeated warning that test automation is a software engineering activity that requires software engineering discipline and adherence to best practices.

STANDARDS AND WHAT AUTOMATION LANGUAGE TO USE...

The test automation implementation language should be in line with what developers use for their implementation tasks. It depends entirely on the project, product, and environment you are working in.

Some guidelines you may want to follow:

---

[29] http://nunit.org/
[30] https://www.soapui.org/
[31] https://www.seleniumhq.org/
[32] http://appium.io/

Standardize on one programming language.

It is a bad idea to have the test engineers work in a programming language (say Python) that is different from what the other software engineers use for the project (say Java). If your project or product is mainly written in Java, have the test engineers use Java as well (meaning have them use the Java language bindings for Selenium). If your product is developed in C#, then have them use C#. The reason for this is threefold:

1. If people assist each other, it is good to have one language as the common base (no Java programmers complaining that they have to look at Python test code).
2. It allows for mobility within the team and encourages people to help each other; folks will not get stuck in one job because they only know a specific scripting language.
3. It makes integration of various tests easier (for example, unit test, web service tests, and functional tests all being written in Java are easier to integrate into the continuous integration setup).

Standardize on one IDE.

Same reasoning as above – you want all team members to work in the same IDE or code editor to minimize friction and allow for flexibility across the team.

Standardize on one desktop / server OS.

Have developers and test engineers work on exactly the same desktop and server OS version. Do not let one developer work on SUSE Linux while the test engineers work on Windows 10. Despite the fact that most tools now claim to be cross-platform, many differences remain. Spare yourself the headache and lots of debugging time.

Standardize on one (set of) browsers.

Make hard choices about what browsers you will support (the product owner needs to make that call). Make sure everybody develops and tests on them (regardless if they are desktop or mobile versions). Be specific down to the version number. Do not let one developer develop their solution on Chrome but expect it to work on IE 11 or Firefox. Keep OS and browser combos in sync. Too many problems crop up with developers developing on Linux / Chrome and expecting things to work on Windows 10 / Edge or MacOS "El Capitan" / Safari 10.

Dictate to your outsourcing partner.

If you outsource test automation development to a partner, you must dictate the automation tool / framework to use; the programming language, desktop / server OS, browsers; and the IDE / code editor, as you will inherit the code, supporting scripts, etc. If you do not do that, do not be surprised if your partner develops in whatever may be convenient to them, but not to you.

Consider enforcing other standards internally and / or with your partner; for example, asking to provide code coverage statistics for each build.

WHAT ABOUT REGRESSION TESTING?

Regression testing is one of those often misunderstood terms that everybody throws around, so here is a brief definition:

- Regression testing is any type of software testing that seeks to **uncover new defects**, or regressions, in existing functionality **after changes have been made to a system**, such as functional enhancements, patches, or configuration changes.
- The intent of regression testing is to **ensure that a change**, such as a software fix, **did not introduce new defects**. One of the main reasons for regression testing is that it is often extremely difficult for a programmer to figure out how a change in one part of the software will echo in other parts of the software.
- Common methods of regression testing include **rerunning previously run tests and checking whether program behavior has changed** and whether previously fixed faults have re-emerged. Regression testing can be used to test a system efficiently by systematically selecting the appropriate minimum set of tests needed to adequately cover a particular change.

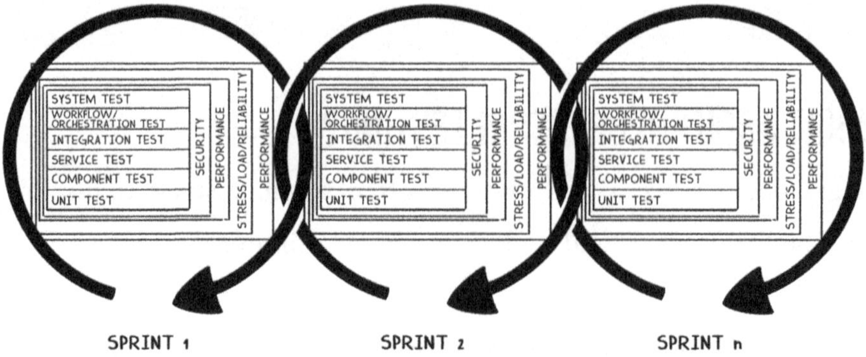

Figure 72 - Iterating over Various Test Categories

Code coverage tools help determine how effective regression tests are. Test automation allows for repeated, cheap, regression test execution cycles.

Without test automation, regression testing quickly leads to tester burnout due to the boring and repetitive nature of rerunning the same tests over and over again.

WHEN TO AUTOMATE FUNCTIONAL TESTING

As mentioned above, risk management and the Pareto analysis should provide your automation targets. Not everything should be automated, as the benefits of complete automation do not outweigh the costs (at least for most companies), and you run the risk of blowing your budget.

Consequently, an 80/20 principle should be applied to determine what functionality can reasonably be automated:

- Basic end-to-end functionality.

- Functionality that is critical to the business or supports many concurrent users (concurrency, stress, load, and redundancy considerations).
- Functionality that has been assigned specific service-level agreements (SLA), with consequences for not meeting them.
- Functionality that touches multiple areas of the application (or other applications).
- Tasks that are either repetitive or error prone, such as data loading, system configuration, etc.
- Testing that involves multiple hardware or software configurations, i.e., the same tests need to be rerun to validate compatibility.

When planning your functional test automation tasks, follow these steps:

1. Review what will be delivered when based on your release / sprint plans.
2. Try to estimate when certain parts of the system will stabilize.
3. Assess each automation target for suitability.
4. Make a clear choice what to automate versus what to leave for manual testing.
5. Document a Definition of Done for you test automation so you know when you have hit your target.

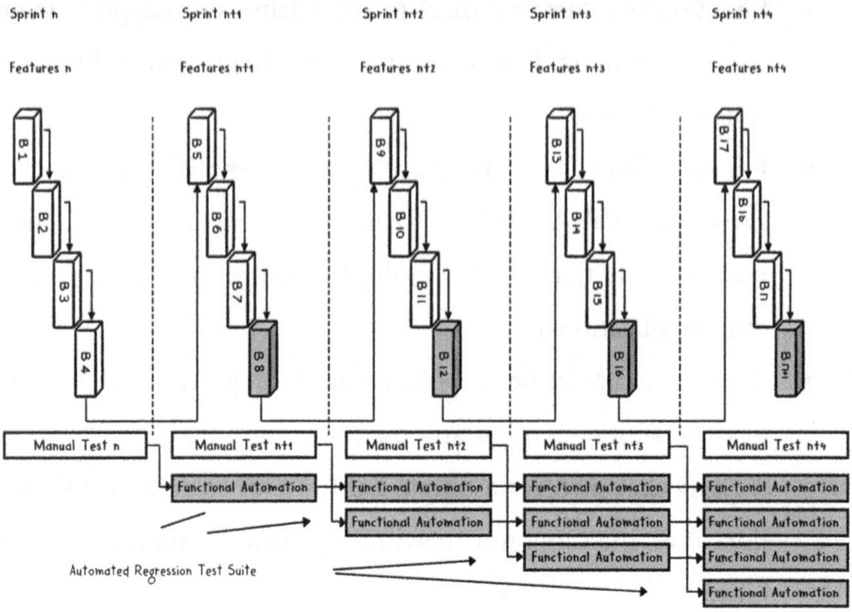

Figure 73 - Automated Regression Test Suite Build Up over Time

HOW UNIT AND FUNCTIONAL AUTOMATION BUILDS UP OVER TIME

Automated unit tests represent a significant amount of your code base.

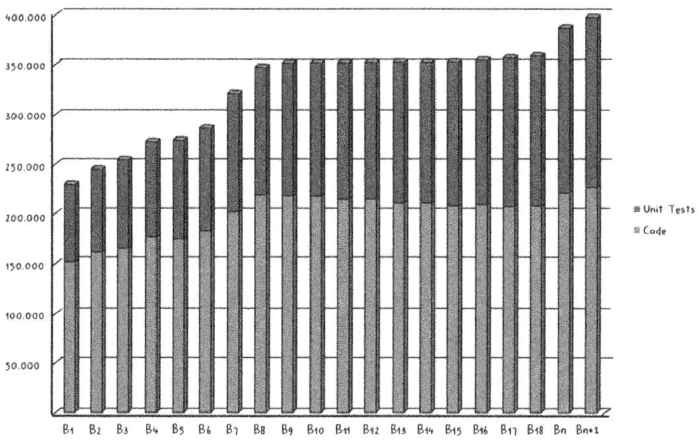

Figure 74 - Automated Unit Test Build Up over Time

Combined with functional tests, test automation gradually builds up over time.

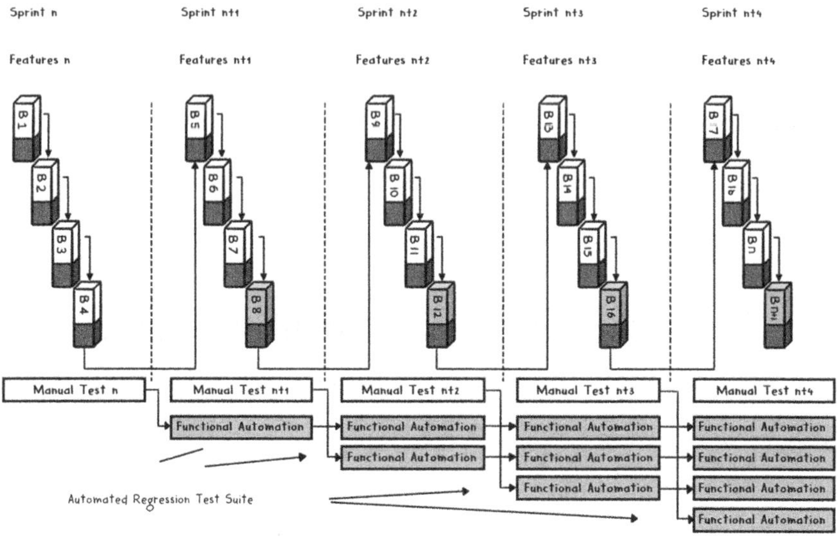

Figure 75 Gradual Test Automation Build Up over Time

Using both automated unit and functional testing provides an effective top-down and bottom-up approach:

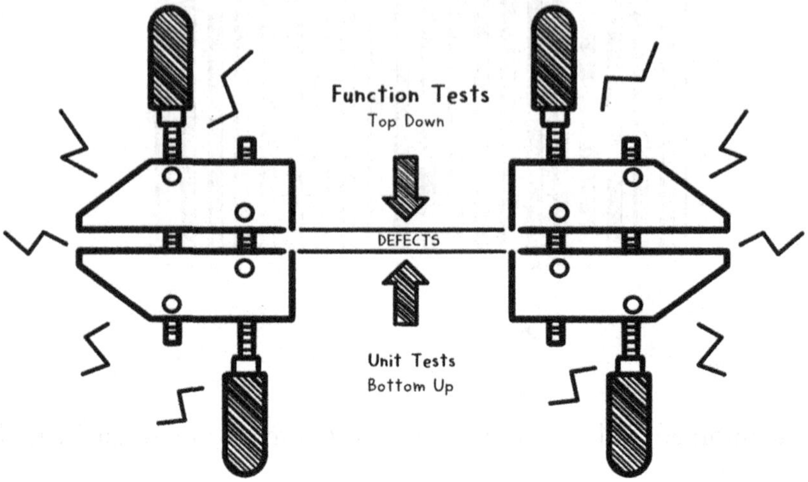

Figure 76 - Top Down and Bottom Up Approach to Test Automation

Having both automated unit tests as well as automated functional tests literally squeezes the defects out of the system. More important, automation provides instant feedback to the developers and hence shortens the time to fix defects.

WE HAVE A LARGE CODE BASE WITH NO UNIT TESTS

It is not uncommon to find big projects with millions of lines of code that have no unit tests. No unit test automation. No continuous integration server. Now what?

Fact is that most projects could never afford to retrofit unit test for a large code base. That is OK; you got to where you are without unit tests.

Now, realizing the benefits, you want to start somewhere:

- You can add unit tests for all new code changes going forward.
- You can add unit tests for all modules / components you are refactoring.
- You can review critical modules / components and decide to add unit tests just to those.
- You can do some defect analysis, determine what modules / components of your system cause the most failures, and invest in adding unit tests to them.

As mentioned in the chapter "***Agile Adoption Pitfalls – PART 7, Ignoring DevOps,***" there is nothing preventing you from using DevOps principles such as automated unit test or continuous integration on any project. You do not have to start a new project. You do not have to be Agile. These principles are universally acknowledged as software engineering best practices.

SYSTEM INSTABILITY AND AUTOMATION

It is not uncommon for complex systems to come together in pieces, meaning different components arrive at different times during the SDLC in various states of "doneness." The trick in any automation effort is to pick the pieces that are stable enough as they are made available in sprints.

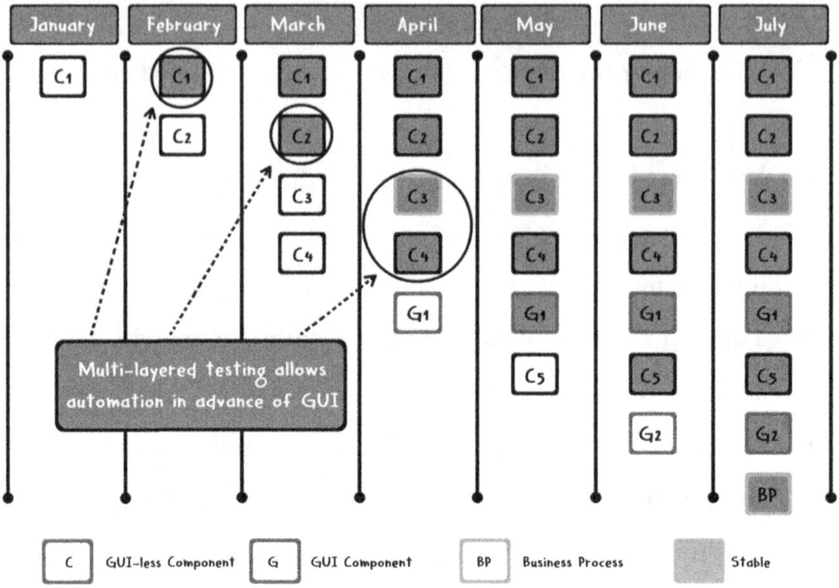

Figure 77 - Progressive Automation as System Components Stabilize

The illustration above shows how you can pick off automation targets as they become available and as they stabilize.

The author has found that it is a useful exercise to visualize your project deliverable like this in order to facility a discussion about suitable automation targets. Most products / projects need to economize and make sure that components are stable enough for automation in order to avoid lots of rework.

WHO SHOULD AUTOMATE WHAT?

Simply put, unit tests should **always** be automated by the developer who wrote the code. The main reason is that the developer is closest to the problem and therefore best qualified to write the unit test.

Some variations apply – for example in **pair programming / XP** or **test-driven development** – but if your project does not follow those, the developer who wrote the code should also write the unit test.

All other types of test automation are up for grabs by the Agile / Scrum team members that are best qualified and have time available. Remember that in Agile / Scrum there is no functional breakdown of job responsibilities. Whoever has time jumps in to help the work cross the finish line. Having said that, the reality is that whoever automates needs decent software engineering skills, including analysis, coding, debugging, and critical thinking skills.

Performance test automation is one specialty that usually requires lower level operating system / database / application server knowledge in order to determine performance bottlenecks.

CODE COVERAGE

The concept of code coverage is based on the structural notion of the code. Code coverage implies a numerical metric that measure the elements of code that have been exercised as a consequence of testing. There are a host of metrics: statements, branches, and data that are implied by the term coverage. There are several well-established tools[33] in the market today that provide this measurement functionality for various kinds of programming environments.

Once you have test automation, at whatever level, these tools will allow you to measure progress from build to build.

---

[33] https://en.wikipedia.org/wiki/List_of_tools_for_static_code_analysis

Integrating automated unit tests, automated functional tests, and code coverage into your continuous integration server will provide you with lots of metrics that will allow for calibration of the development effort. If you want to improve something, measure it!

TEST AUTOMATION ROI CALCULATIONS

This section deals with **return on investment (ROI)** calculations for test automation efforts. Nine out of 10 companies that consider automation do not really understand the potential benefits of automation (the return), or the cost involved.

According to Wikipedia,[34] "*the purpose of the "return on investment" (ROI) metric is to measure, per period, rates of return on money invested in an economic entity in order to decide whether or not to undertake an investment. It is also used as indicator to compare different project investments within a project portfolio. The project with best ROI is prioritized.*"

The simplest ROI calculation looks like this:

$$ROI = \frac{\text{Gains} - \text{Investment Costs}}{\text{Investment Cost}}$$

When considering the "business case" for test automation, there are three approaches for calculating ROI:

1. Basic ROI calculation
2. Efficiency ROI calculation

[34] https://en.wikipedia.org/wiki/Return_on_investment

3. Life insurance ROI calculation

Basic ROI

The **basic ROI calculation** focuses on an "apples to oranges" comparison of manual test execution versus automated test execution. The assumption is that automated testing will reduce testing cost either by replacing manual testing substantially or by supplementing it. This calculation is useful when you make tradeoff decisions about hiring additional testers or trying to automate part of the testing process.

The investment cost includes such automation costs as hardware, software licenses, hosting fees, training, automation development and maintenance, and test execution and analysis.

The gain is set equal to the pre-automation cost of executing the same set of tests.

Sample Calculation

Let's say we're working on a project with an application that has several test cycles that equates to a weekly build 6 months out of the year. In addition, the project has the following manual and automation parameters that will be used in calculating ROI:

General Parameters

- 1,500 test cases, 1,000 of which can be automated
- Tester hourly rate – $45 per hour

Manual Parameters

- Manual test execution/analysis time (average per test) – 10 minutes

Automation Parameters

- Tool and license cost (5 licenses) – $20,000
- Tool training cost – $5,000
- Test machine cost (3 machines) – $3,000
- Test development / debugging time (average per test) – 60 minutes (1 hour)
- Test execution time – 2 minutes
- Test analysis time – 4 hours for 500 tests
- Test maintenance time– 8 hours per build

In order to calculate the ROI, we need to calculate the **investment** and the **gains**. The investment costs can be calculated by converting automation factors to dollar amounts. The automated tool and licensing costs, training costs, and machine costs stand on their own, but the other parameters are calculated as follows:

- The **automated test development time** can be converted to a dollar figure by multiplying the average hourly automation time per test (1 hour) by the number of tests (1,000), then by the tester hourly rate ($45). **1,000 x 45 = $45,000**
- The **automated test execution time** doesn't need to be converted to a dollar figure in this example, because the tests will ideally run independently on one of the automated test machines.

- The **automated test analysis time** can be converted to a dollar figure by multiplying the test analysis time (4 hours per week, given that there is a build once a week) by the timeframe being used for the ROI calculation (6 months or approximately 24 weeks), then by the tester hourly rate ($45). **4 x 24 x 45 = $4,320**
- The **automated test maintenance time** can be converted to a dollar figure by multiplying the maintenance time (8 hours per week) by the timeframe being used for the ROI calculation (6 months or approximately 24 weeks), then by the tester hourly rate ($45). **8 x 24 x 45 = $8,640**

The total **investment cost** can now be calculated as:
$20,000 + $5,000 + $3, 000 + $45,000 + $4,320 + $8,640 = **$85,960**

The gain can be calculated in terms of the manual test execution / analysis time that will no longer exist once the set of tests has been automated. The manual execution / analysis time can be converted to a dollar figure by multiplying the execution / analysis time (10 minutes or 0.17 hours) by the number of tests (1,000), then by the timeframe being used for the ROI calculation (6 months or approximately 24 weeks), and finally by the tester hourly rate ($45).

Manual test development and maintenance are not considered because these activities must take place regardless of whether or not a test is automated.

The **gain** is therefore:
0.17 x 1,000 x 24 x $45 = **$183,600**

$$ROI = \frac{\text{Gains} - \text{Investment Costs}}{\text{Investment Cost}} = \frac{\$183{,}600 - \$85{,}960}{\$85{,}960} = 113.6\%$$

The ROI is calculated at 113.6%. Note that over time this ROI percentage will increase, because the tool costs eventually get replaced in the calculation by tool support costs.

Pros and Cons

The advantage to using this **basic ROI calculation** is that the project dollar figure makes it good for communication with upper level management.

The main drawback is that it sends an overly simplistic message and makes people think that automation will replace testers. It assumes that the automated tests completely replace their manual counterparts, but that is almost never the case. Manual exploratory testing is still necessary, especially until the system / component is stable enough for automation.

On the other hand, automation, once working, allows for an almost infinite number of variations. For example, whereas a tester might only test the upper and lower boundary value of a data element in order to save time, automation can easily run through 500 or 5,000 values.

The final fallacy is that often calculations like this assume a fixed set of tests. The fact is that as you automate things and they run without human intervention, testers that are freed up will refocus on other areas of your application, so the project budget seldom decreases due to automation. Funds are usually redistributed for better testing on other parts of the application.

Efficiency ROI

The **efficiency ROI calculation** is based on the **basic ROI calculation**, but only considers the time investment gains for assessing testing efficiency, as opposed to looking at monetary values. This calculation is useful for projects that have had an automated tool for a long enough time that there isn't much need to give much consideration to its cost in the ROI calculation. In addition, it may be better suited for calculation by test engineers, because the factors used are easily attainable.

Sample Calculation

The project example from the **basic ROI calculation** will be used for this sample calculation also. The main difference is that the investment and gains will need to be calculated considering the entire bed of functional tests, not just the ones that are automated.

The investment cost is derived by calculating the time investment required for automation development, execution, and analysis of 1,000 tests, and then adding the time investment required for manually executing the remaining 500 tests.

Calculations need to be done in terms of days as opposed hours, because the automated tests are ideally operated in 24-hour days, while manual tests operate in 8-hour days. Since tests runs can often abruptly stop during overnight runs, however, it is usually a good practice to reduce the 24-hour-day factor to a more conservative estimate of about 18 hours.

- The **automated test development time** is calculated by multiplying the average hourly automation time per test (1 hour) by the number of tests (1,000), then dividing by 8 hours to convert the figure to days. (1 x 1,000) / 8 = **125 days**
- The **automated test execution time** must be calculated in this example, because time is instrumental in determining the test efficiency. This is calculated by multiplying the automated test execution time (2 min or 0.03 hours) by the number of tests per week (1,000), by the timeframe being used for the ROI calculation (6 months or approximately 24 weeks) then dividing by 18 hours to convert the figure to days. (0.03 x 1,000 x 24) /18 = **40 days**. (This number will be reduced when tests are split up and executed on different machines, but for simplicity we will use the single machine calculation.)
- The **automated test analysis time** can be calculated by multiplying the test analysis time (4 hours per week given that there is a build once a week) by the timeframe being used for the ROI calculation (6 months or approximately 24 weeks), then dividing by 8 hours (since the analysis is still a manual effort) to convert the figure to days. (4 x 24) / 8 = **12 days**
- The **automated test maintenance time** is calculated by multiplying the maintenance time (8 hours) by the timeframe being used for the ROI calculation (6 months or approximately 24 weeks), then dividing by 8 hours (since the maintenance is still

a manual effort) to convert the figure to days. (8 x 24) / 8 = **24 days**

- The **manual execution time** is calculated by multiplying the manual test execution time (10 min or 0.17 hours) by the remaining manual tests (500), then by the timeframe being used for the ROI calculation (6 months or approximately 24 weeks), then dividing by 8 to convert the figure to days. 0.17 x 500 x 24 /8 = **255 days** (Note that this number is reduced when tests are split up and executed by multiple testers, but for simplicity we will use the single-tester calculation.)

The total time investment can now be calculated as:

125 + 40 + 12 + 24 + 255 = **456 days**

Now turning our attention to the gain, you'll find that it can be calculated in terms of the manual test execution / analysis time.

The **manual execution / analysis time** can be converted to a dollar figure by multiplying the execution / analysis time (10 minutes or 0.17 hours) by the total number of tests (1,500), then by the timeframe being used for the ROI calculation (6 months or approximately 24 weeks), then dividing by 8 hours to convert the figure to days. (0.17 x 1,500 x 24) / 8 = **765 days** (Note that this number is reduced when tests are split up and executed by multiple testers [which would have to be done in order to finish execution within a week], but for simplicity we will use the single-tester calculation.)

Inserting the **investment** and **gains** into our formula:

$$ROI = \frac{\text{Gains} - \text{Investment Costs}}{\text{Investment Cost}} = \frac{765 - 456}{456} = 67.8\%$$

The ROI is calculated at 67.8%.

Pros and Cons

As discussed in the **basic ROI calculation** section, project dollar figures are seldom reduced due to automation, so using the **efficiency calculation** allows focus to be removed from potentially misleading dollar amounts. In addition, it provides a simple ROI formula that doesn't require sensitive information such as tester hourly rates.

This therefore makes it easier for test engineers and lower-level management to present the benefits of test automation when asked to do so.

Many of the other cons still exist, however:

- This calculation still maintains the assumption that the automated tests completely replace their manual counterparts – a common fallacy which is worth not repeating or reinforcing.
- It assumes that full regression is done during every build even without test automation. This may not be the case, however. Full regression may not be performed without test automation, so introducing test automation increases coverage and reduces project risks as opposed to increasing efficiency.

Life Insurance ROI

The **life insurance ROI calculation** seeks to address ROI concerns left by other calculations by looking at automation benefits independently of manual testing. Test automation is not intended to replace manual testing.

Really the best way to look at automation is to acknowledge that not everything can or should be automated but that automation is necessary to:

- Provide near real-time feedback to development
- Lower cycle times and increase repeatability
- Improve efficiency
- Enable automated metric gathering via code coverage
- Provide insurance against catastrophic failure

As pointed out in the section about **risk management and prioritization**, focusing on automation using 80/20 analysis helps optimize work. Test automation saves time in test execution, which provides testing resources with more time for increased analysis, test design, development, and execution of new tests. Test automation is also a foundational best practice when trying to follow DevOps principles and automate the software delivery pipeline.

It also provides more time for ad hoc and exploratory testing. This equates to increased coverage, which reduces the risk of production failures. By assessing the risk of not performing automation (relative to potential production failures), and calculating the cost to the project if the risk turns into a loss, this calculation addresses the ROI relative to an increased quality of testing. Basically this is a life insurance calculation.

Sample Calculation

The project example from the **basic ROI calculation** will be used for this sample calculation also, and the investment costs are calculated the exact same way. The investment cost is therefore **$85,960**.

The gain is calculated in terms of the amount of money that would be lost if a production defect was discovered in an untested area of the application. The loss may relate to production support and application fixes, and may also relate to lost revenue due to users abandoning the application or just not having the ability to perform the desired functions in the application. The potential loss is assumed at $450,000.

Inserting the **investment** and **gains** into our formula:

$$ROI = \frac{\text{Gains} - \text{Investment Costs}}{\text{Investment Cost}} = \frac{\$450{,}000 - \$85{,}960}{\$85{,}960} = 423.5\%$$

The ROI is calculated at 423.5%.

Pros and Cons

The advantage to using this calculation is that it addresses some of the issues left by other calculations by eliminating the comparison between automated and manual test procedures, and dealing with the positive effect of increased test coverage.

The **risk reduction ROI** is not without its own flaws, however. One is that it relies heavily on subjective information.

Loss calculations are not absolutes, because no one knows for sure how much money could be lost. It could be much more than what's estimated, or it could be much less. In addition, it requires a high degree of risk analysis and loss calculations.

In real world projects, it is often difficult to get anyone to identify and acknowledge risks, much less ranking risks, and calculating potential loss. One final disadvantage is that without the manual to automation comparison, it still leaves the door open for people to ask, "why not just allocate more resources for manual testing instead of automated testing?"

Ways to Improve Test Automation ROI

There are several common factors that affect each of the ROI calculations, so improvements to any of those factors will certainly improve ROI, regardless of how it's calculated. These factors are as follows:

- Automated test development / debugging time (average per test)
- Automated test execution time
- Automated test analysis time
- Automated test maintenance time

Automated test development / debugging can be decreased by hiring well-qualified software engineers with the right skill sets as well as using industry-acknowledged best practice tools with established track records (such as the aforementioned xUnit frameworks or Selenium).

Automated test execution time can be decreased by better exception handling routines. In addition, automated test failures often compound upon one another, so following clean design principles that make tests independent of each other is essential.

Automated test analysis time can be decreased by building better reporting mechanisms into the tests.

Automated test maintenance time may be decreased by carefully assessing when application functionality is ready for test automation (see "System Instability and Automation"). Additionally, increasing modularity and adding comments to automated tests is another way to make maintenance easier and faster. All common software engineering best practices apply to test automation code as well.

Part II

Agile Adoption Challenges

CHAPTER 18

AGILE ASSESSMENTS – HOW TO ASSESS AND EVALUATE AGILE / SCRUM PROJECTS

A common problem many Agile / Scrum adopters have is how to assess and evaluate their projects. This situation arises when auditing organizations transitioning from waterfall approaches, as well as organizations that wish to improve existing practices along the Agile / Scrum value chain.

We have all seen this situation before: You argued for Agile / Scrum; you convinced your supervisors, executives, and peers; you justified the training and transition cost; you retooled and adjusted current processes; you hired consultants and new employees experienced in Agile / Scrum, etc.

But, maybe after the first release of your product or after the tenth project wrapped up, somebody is going to ask the inevitable questions:

- *"Are we doing better than before?"*
- *"How are our projects doing?"*
- *"Are we doing better than last year?"*

These kinds of questions are healthy. Simply adopting an Agile / Scrum process will not guarantee that you actually deliver software better or faster.

Because you want to continuously get better at what you are doing, you need to know what areas to focus on. It is important to understand key improvement areas for the next time around or the next budget cycle. With yearly budgets, you need to know where to spend your money wisely, so you spend it where it counts most.

There are three times in an Agile adoption process you might want to ask these kinds of questions:

1. After the initial Agile / Scrum pilot project has finished – to assess how well that pilot project performed.
2. After the rollout of the Agile / Scrum process to multiple projects or across the organization – to assess how projects are doing compared with each other.

3. At regular intervals for each project that adopted the Agile / Scrum process – to assess key improvement targets on a regular basis.

But even if you already adopted Agile / Scrum years ago, measuring where you stand will help you assess where you are and where you need to go – so regardless, going through an assessment is a great exercise because it summarizes and visualizes where you stand.

How to go about an Agile / Scrum Project Assessment

The thousand-mile-high view is that you want to get to a radar chart that shows key assessment areas like this:

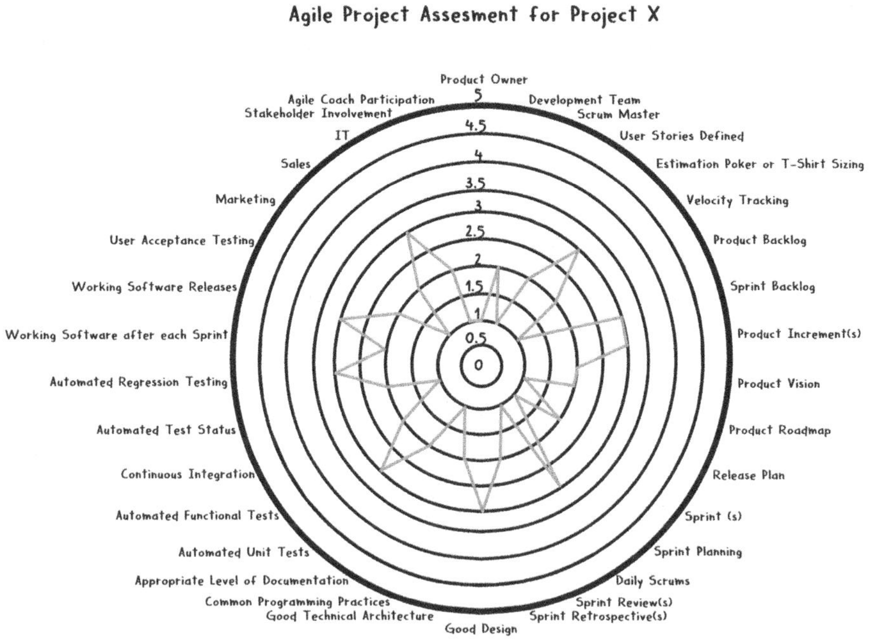

Figure 78 - Agile / Scrum Assessment for one project

This chart simply shows 34 assessment items (organized in 7 assessment areas) on an assessment rating scale of 1 (worst) to 5 (best). The assessment rating scale is defined as follows:

- 5 = Consistently followed in the spirit of Agile / Scrum development; fully implemented role, artifact, process, or best practice the way it is intended in Agile / Scrum.
- 4 = More of less consistently followed; mostly implemented role, artifact, process, or best practice.
- 3 = Inconsistently followed or done incorrectly; role, artifact, process, or best practice is inconsistently implemented or followed, or used incorrectly (for example, calling regular meet-

ing "Daily Scrums").

- 2 = Rarely followed.
- 1 = Never followed / not adhered to / not implemented.

The blue line indicates the area that would have been covered if every assessment area were a solid 5. The orange line shows the area representative for the actual sample project.

This kind of radar chart is simple enough to generate in Microsoft Excel. Here is the example data used for the radar chart above.

#	Area	#	Item	Target Value	Current Value
1	Roles & Responsibilities	1	Product Owner	5	1
		2	Development Team	5	2
		3	Scrum Master	5	1
2	Planing & Estimation	4	User Stories Defined	5	2
		5	Estimation Poker or T-Shirt Sizing	5	3
		6	Velocity Tracking	5	2
3	Artifacts	7	Product Backlog	5	1
		8	Sprint Backlog	5	3
		9	Product Increment(s)	5	3
		10	Product Vision	5	2
		11	Product Roadmap	5	2
		12	Release Plan	5	1
4	Process-Scrum	13	Sprint(s)	5	2
		14	Sprint Planning	5	1
		15	Daily Scrums	5	3
		16	Sprint Review(s)	5	1
		17	Sprint Retrospective(s)	5	2
5	Process-Engineering Best Practices	18	Good Design	5	3
		19	Good Technical Architect	5	2
		20	Common Programming Practices	5	1
		21	Appropriate Level of Documentation	5	2
		22	Automated Unit Tests	5	3
		23	Automated Functional Test	5	2
		24	Continuous Integration	5	1
		25	Automated Test Status	5	2
		26	Automated Regression Testing	5	3
6	Delivery Effectiveness	27	Working Software After Each Sprint	5	2
		28	Working Software Releases	5	3
		29	User Acceptance Testing	5	2
7	Organizational Support	30	Marketing	5	1
		31	Sales	5	2
		32	IT	5	3
		33	Stakeholder Involvement	5	2
		34	Agile Coach Participation	5	1

Table 12 - Agile / Scrum Assessment Areas

The assessment areas and items in this table are a good starting point, but there is no reason why you could not become more detailed and add additional areas or items as you see fit. Or, if you only want to focus in on Agile / Scrum process issues, you could delete some areas and items. This is simply a baseline for you to take and adapt as needed.

The following section will briefly review each of the assessment areas and items in turn.

ROLES & RESPONSIBILITIES

Are the roles and responsibilities clearly defined and filled with quality personnel, according to Agile / Scrum best practices?

Product Owner

- Single person responsible for maximizing the return on investment (ROI) of the development effort
- Responsible for product vision
- Constantly re-prioritizes the product backlog, adjusting any long-term expectations such as release plans
- Final arbiter of requirements questions
- Accepts or rejects each product increment
- Decides whether to ship
- Decides whether to continue development
- Considers stakeholder interests
- May contribute as a team member
- Has a leadership role

Development Team

- Cross-functional (e.g., includes members with testing skills, and often others not traditionally called developers: business analysts, domain experts, etc.)

- Self-organizing / self-managing, without externally assigned roles
- Negotiates commitments with the product owner, one sprint at a time
- Has autonomy regarding how to reach commitments
- Intensely collaborative
- Most successful when located in one team room, particularly for the first few sprints
- Most successful with long-term, full-time membership
- 7 +/- 2 members

Scrum Master

- Facilitates the Scrum process
- Helps resolve impediments
- Creates an environment conducive to team self-organization
- Captures empirical data to adjust forecasts
- Shields the team from external interference and distractions
- Enforces time boxes
- Keeps Scrum artifacts visible
- Promotes improved engineering practices
- Has no management authority over the team
- Has a leadership role

Planning and Estimation

Are Agile / Scrum best practices used for planning and estimation activities?

User Stories Defined

Are user stories defined? Are they granular enough? Are the defined according to the **INVEST** principle?

- **Independent** (to the degree possible), i.e., does not need other User Stories to be implemented
- **Negotiable**
- **Valuable**, focused on features, not tasks; written in the user's language
- **Estimable**
- **Small**, half a Sprint or less for one development team member
- **Testable**

Estimation Poker or T-Shirt Sizing

Do all team members understand the difference between relative and absolute estimation techniques?

T-Shirt Sizing

- XS = 1 point
- Small = 2 points
- Medium = 3 points
- Large = 5 points

- XL = 8 points
- XXL = 13 points
- EPIC = product owner needs to break down

Estimation Poker

- Based on team votes and consensus
- A relative estimation technique
- Accuracy over precision
- Based on the Fibonacci number sequence (1, 2, 3, 5, 8, 13, 21, 34, 55, ...)
- Done by the development team members doing the work
- Self-corrects for team idiosyncrasies
- Enables more accurate long-term planning

Velocity Tracking

Do you understand your team's performance? Is it 36 story points or 56 on average?

How to Calculate

- Rolling average of sprints
- Maximum
- Minimum
- Lifetime average

How to Use

- To determine sprint scope
- To calculate approximate cost of a release
- To track release progress

Remember Best Practices

- Velocity without a quality metric is not useful
- Velocity of two teams is not comparable
- Velocity changes when the team composition changes
- Velocity increases with team's tenure
- Velocity does not equal productivity
- Defects and rejected user stories count against velocity!

ARTIFACTS

Are best practices followed for all relevant Agile / Scrum artifacts?

Product Backlog

- Force-ranked list of desired functionality
- Visible to all stakeholders
- Any stakeholder (including the development team) can add items
- Constantly re-prioritized by the product owner
- Items at top are more granular than items at bottom
- Maintained during the backlog refinement meeting

Sprint Backlog

- Consists of committed product backlog items negotiated

between the team and the product owner during the sprint planning meeting
- Scope commitment is fixed during sprint execution
- Initial tasks are identified by the team during sprint planning meeting
- Team will discover additional tasks needed to meet the fixed scope commitment during sprint execution
- Visible to the team
- Referenced during the daily scrum meeting

Product Increments

- Within the context of a single sprint, an increment is a work product that has been done and is deemed "shippable" or "deployable"
- Developed
- Tested
- Integrated
- Documented
- An increment may, or may not, be released externally depending on the release plan

Product Vision

Does a **product vision** exist? The product owner or the product management team identifies the product vision. The product vision is a definition of what your product is, how it will support your company or organizations' strategy, and who will use the product. On longer projects, the product vision is revisited at least once a year.

Product Roadmap

Does the project / product have a **product roadmap**? The product roadmap is a high-level view of the product requirements, with a loose time frame for when you will develop those requirements. Identifying product requirements and then prioritizing and roughly estimating the effort for those requirements is a large part of creating your product roadmap. On longer projects, the product roadmap is revisited at least twice a year.

Release Plan

Is there a **release plan**? The release plan identifies a high-level timetable for the release of working software. An Agile project will have many releases, with the highest-priority features launching first. A typical release includes three to five sprints. The release plan is created at the beginning of each release.

PROCESS – FOR EXAMPLE, SCRUM

Scrum is a management framework for incremental product development using one or more cross-functional, self-organizing teams of about seven people each. It provides a structure of roles, meetings, rules, and artifacts. Teams are responsible for creating and adapting their processes within this framework.

Sprint(s)

Does the team follow fixed duration sprints? Scrum uses fixed-length iterations, called **sprints**, which are typically two weeks or 30 days long. Scrum teams attempt to build a potentially shippable (properly tested) product increment every iteration.

Sprint Planning

Does the team conduct **sprint planning meetings**? At the beginning of each sprint, the product owner and the team hold a sprint planning meeting to negotiate which product backlog items they will attempt to convert to working product during the sprint. The product owner is responsible for declaring which items are the most important to the business. The team is responsible for selecting the amount of work they feel they can implement without accruing technical debt. The team "pulls" work from the product backlog to the sprint backlog.

Daily Scrums

Does the team conduct **daily scrum meetings**? Every day at the same time and place, the scrum development team members spend a total of 15 to 30 minutes reporting to each other. Each team member summarizes what he did the previous day, what he will do today, and what impediments he faces. Standing up at the daily scrum will help keep it short. Topics that require additional attention may be discussed by whoever is interested after every team member has reported.

Sprint Review(s)

Does the team conduct **sprint reviews**? After sprint execution, the team holds a sprint review meeting to demonstrate a working product increment to the product owner and everyone else who is interested. The meeting should feature a live demonstration, not a report.

After the demonstration, the product owner reviews the commitments made at the sprint planning meeting and declares which items are now considered done. For example, a software item that is merely "code complete" is considered not done, because untested software is not shippable.

Incomplete items are returned to the product backlog and ranked according to the product owner's revised priorities as candidates for future sprints.

Sprint Retrospective (s)

Does the team conduct **sprint retrospectives**? Each sprint ends with a retrospective. At this meeting, the team reflects on its own process. They inspect their behavior and take action to adapt it for future sprints. Dedicated scrum masters will find alternatives to the stale, fearful meetings everyone has come to expect.

An in-depth retrospective requires an environment of psychological safety not found in most organizations. Without safety, the retrospective discussion will either avoid the uncomfortable issues or deteriorate into blaming and hostility.

PROCESS – ENGINEERING BEST PRACTICES

Is your project utilizing industry best practice knowledge of what works and what does not work?

Good Design & Good Technical Architecture

Depending on your technology stack, you must pick and choose the appropriate architecture for your application. Component technologies, web services, and commonly used design patterns all provide best practice guidance for your application.

Common Programming Practices

Does your team follow common programming practices such as following a coding standard, peer reviews, code review, and implementing unit tests? Do you use source control effectively?

Appropriate Level of Documentation

Are you following the appropriate level of documentation? There are different documentation requirements between a small, internally used, application versus a mission / life critical application made up of millions of lines of code.

Automated Unit Tests

Has your team implemented automated unit tests using a framework like nUnit?

Automated Functional Tests

Has your team automated functional tests? Automated testing tools (such as Selenium) are capable of executing tests, reporting outcomes, and comparing results with earlier test runs. Tests carried out with these tools can be run repeatedly, at any time of day. The method or process being used to implement automation is called a test automation framework.

Continuous Integration

Does your team utilize continuous integration (CI)? A best practice for constructing code includes the daily build and smoke test. Martin Fowler goes further and proposes continuous integration that also integrates the concept of unit tests and self-testing code. Kent Beck supposedly said that "daily builds are for wimps," meaning only continuous integration would provide the best benefits from a coding perspective.

Ultimately the frequency needs to be decided by the team, but the more frequent, the better. Even though continuous integration and unit tests have gained popularity through XP, you can use these best practices on all types of projects. Standard frameworks to automate builds and testing, such as Jenkins, Hudson, Ant, and xUnit, are freely available, and new open source tools appear on a regular basis.

Automated Test Status

Is test status automatically generated from all kinds of automated testing (unit, functional) so test status is available and visible to everyone?

Automated Regression Testing

Are automated regression tests executed with every build? Regression testing is one of those often misunderstood terms that everybody throws around, so here is a brief definition:

- Regression testing is any type of software testing that seeks to uncover new defects, or regressions, in existing functionality after changes have been made to a system, such as functional enhancements, patches, or configuration changes.
- The intent of regression testing is to ensure that a change, such as a software fix, did not introduce new defects. One of the main reasons for regression testing is that it is often extremely difficult for a programmer to figure out how a change in one part of the software will echo in other parts of the software.
- Common methods of regression testing include rerunning previously run tests and checking whether program behavior has changed and whether previously fixed faults have reemerged. Regression testing can be used to test a system efficiently by systematically selecting the appropriate minimum set of tests needed to adequately cover a particular change.

Code coverage tools help determine how effective regression tests are. Test automation allows for repeated, cheap, regression test execution cycles.

Without test automation, regression testing quickly leads to tester burnout due to the boring and repetitive nature of rerunning the same tests over and over again.

DELIVERY EFFECTIVENESS

Is your team delivering working software on time according to the expectations of the product owner and end users?

Working Software after each Sprint

After each sprint, is the software "usable"? Can you demo it? Can you interact with it?

Working Software Releases

After a series of sprints, does your team deliver a software release that works?

User Acceptance Testing

If you are delivering to internal customers, do you perform user acceptance testing? If you are delivering to external customers, do you perform external pre-release or beta testing?

ORGANIZATIONAL SUPPORT

Does your Agile / Scrum project have organizational support?

Marketing

If delivering to the market, does your marketing team support your

Agile / Scrum project? Are there marketing plans? Advertising? Trade shows? How do you plan to gain market traction?

Sales

If your Agile / Scrum project is supposed to be sold by a professional sales organization, do your sales people know how to sell it? What are the sales targets? If you are selling through software downloads or through app stores (iOS / Android), how are you going to attract "eyeballs"?

IT

Is IT prepared to support your Agile / Scrum project once it is released and goes to market? Are your production servers set up and configured, backups scheduled / verified? In short, is your production system operational?

Stakeholder Involvement

Are your stakeholders aware of your Agile / Scrum project? Did they provide feedback? Are they engaged? Involved stakeholders ensure that you do not run into political headwinds and avoid surprises.

Agile Coach Participation

An **agile coach** can provide valuable insight that helps your team be more productive:

- Provide an outside view – coaches bring a new perspective to the organization, team, and individuals, and they remove intrinsic bias and interpersonal issues.

- Identify opportunities for improvement – a good coach provides learning and training for the employees and develops their skills.
- Expose challenges – agile coaching creates an environment that allows teams to face the difficulties and work on a proper solution instead of sweeping problems under the table.
- Save time and money – coaches bring both the tried and the new practices and processes to a team and organization, reducing the degree of trial and error commonly found in homegrown experimentation.

ASSESSING MULTIPLE PROJECTS

Multiple projects can be assessed in very much the same way. Assessing each individual project and comparing their values shows where one project team is performing better in certain areas than others.

This is important when assessing company-wide capabilities or identifying key improvement areas across multiple projects.

The following table and radar graph shows a sample comparison of 5 projects:

#	Area	#	Item	Target6 Value	Project A	Project B	Project C	Project D	Project E
1	Roles & Responsibilities	1	Product Owner	5	1	1	1	1	1
		2	Development Team	5	2	2	5	4	1
		3	Scrum Master	5	3	1	1	1	1
2	Planing & Estimation	4	User Stories Defined	5	2	2	2	2	2
		5	Estimation Poker or T-Shirt Sizing	5	3	3	5	4	2
		6	Velocity Tracking	5	3	1	2	2	2
3	Artifacts	7	Product Backlog	5	1	1	1	1	1
		8	Sprint Backlog	5	3	3	3	3	3
		9	Product Increment(s)	5	3	3	3	4	2
		10	Product Vision	5	2	2	2	2	2
		11	Product Roadmap	5	2	2	2	2	2
		12	Release Plan	5	3	1	1	1	1
4	Process-Scrum	13	Sprint(s)	5	2	2	2	2	2
		14	Sprint Planning	5	1	1	1	1	1
		15	Daily Scrums	5	3	3	5	4	2
		16	Sprint Review(s)	5	1	1	1	1	1
		17	Sprint Retrospective(s)	5	3	1	2	2	2
5	Process-Engineering Best Practices	18	Good Design	5	3	3	3	3	3
		19	Good Technical Architect	5	2	2	2	2	2
		20	Common Programming Practices	5	1	1	1	1	1
		21	Appropriate Level of Documentation	5	2	2	2	2	1
		22	Automated Unit Tests	5	3	3	5	4	3
		23	Automated Functional Test	5	2	2	2	2	2
		24	Continuous Integration	5	1	1	1	1	1
		25	Automated Test Status	5	2	2	2	2	2
		26	Automated Regression Testing	5	3	1	3	3	3
6	Delivery Effectiveness	27	Working Software After Each Sprint	5	2	2	2	2	2
		28	Working Software Releases	5	3	3	5	4	1
		29	User Acceptance Testing	5	3	1	2	2	2
7	Organizational Support	30	Marketing	5	1	1	1	1	1
		31	Sales	5	2	2	2	2	2
		32	IT	5	3	3	5	4	1
		33	Stakeholder Involvement	5	2	2	2	2	2
		34	Agile Coach Participation	5	3	1	1	1	1

Table 13 - Assessment comparing several projects

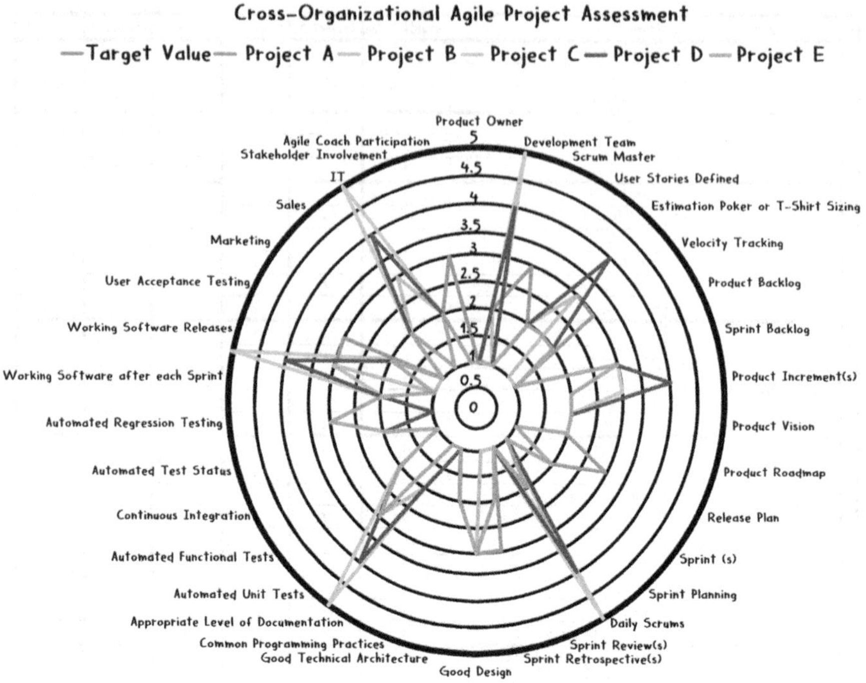

Figure 79 - Assessment comparing several projects

Mapping Project Improvement Targets

Using the same approach, improvement targets can be identified and visualized, as shown in this example:

#	Area	#	Item	Target6 Value	Dec-2014	Jun-2015	Dec-2015	jun-2016	Dec-2016
1	Roles & Responsibilities	1	Product Owner	5	1	1	2	2	2
		2	Development Team	5	2	2	3	3	3
		3	Scrum Master	5	3	3	3	3	3
2	Planing & Estimation	4	User Stories Defined	5	2	2	2	2	2
		5	Estimation Poker or T-Shirt Sizing	5	3	3	3	3	4
		6	Velocity Tracking	5	3	3	3	3	3
3	Artifacts	7	Product Backlog	5	4	4	4	5	5
		8	Sprint Backlog	5	3	4	4	5	5
		9	Product Increment(s)	5	3	3	3	4	4
		10	Product Vision	5	2	2	2	2	2
		11	Product Roadmap	5	2	2	2	2	2
		12	Release Plan	5	1	1	1	1	1
4	Process-Scrum	13	Sprint(s)	5	2	3	4	5	5
		14	Sprint Planning	5	1	2	3	4	4
		15	Daily Scrums	5	3	3	5	5	5
		16	Sprint Review(s)	5	1	3	4	5	5
		17	Sprint Retrospective(s)	5	3	5	5	5	5
5	Process-Engineering Best Practices	18	Good Design	5	3	3	3	3	3
		19	Good Technical Architect	5	2	2	2	2	3
		20	Common Programming Practices	5	1	1	1	1	1
		21	Appropriate Level of Documentation	5	2	2	2	2	1
		22	Automated Unit Tests	5	1	1	2	2	3
		23	Automated Functional Test	5	1	2	3	4	4
		24	Continuous Integration	5	1	1	5	5	5
		25	Automated Test Status	5	1	1	1	1	2
		26	Automated Regression Testing	5	1	1	3	3	4
6	Delivery Effectiveness	27	Working Software After Each Sprint	5	3	3	4	4	5
		28	Working Software Releases	5	3	3	4	4	5
		29	User Acceptance Testing	5	3	3	3	3	3
7	Organizational Support	30	Marketing	5	1	2	3	4	5
		31	Sales	5	2	2	3	4	5
		32	IT	5	2	2	3	4	5
		33	Stakeholder Involvement	5	2	2	3	4	5
		34	Agile Coach Participation	5	1	1	5	5	5

Table 14 - Assessments tracking improvements over time

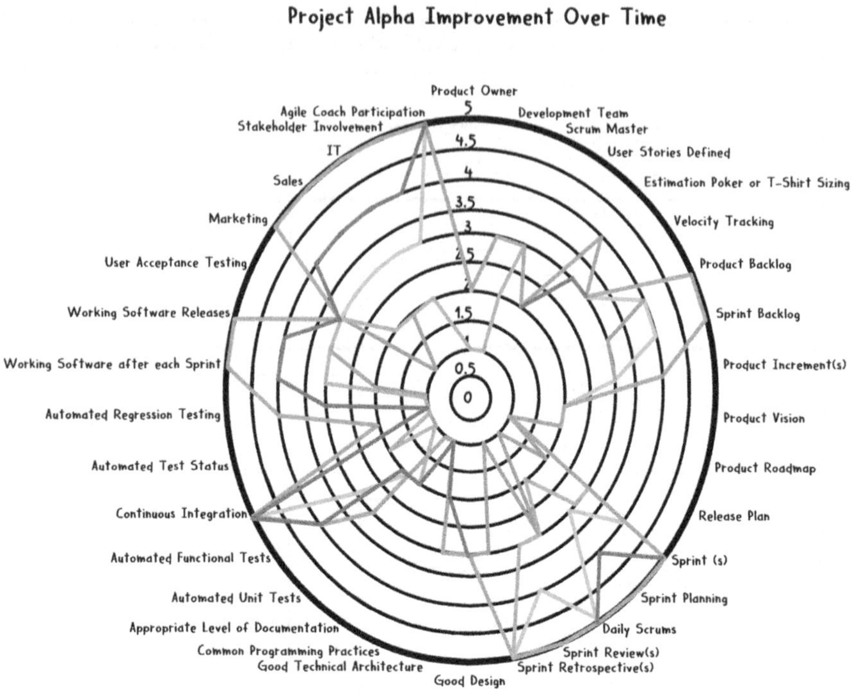

Figure 80 - Assessments tracking improvements over time

SMALL ASSESSMENTS VERSUS ENTERPRISE ASSESSMENTS

Small, single-project, assessments can be done fairly quickly and informally. Small projects can be assessed in a matter of hours. Usually team members know exactly what is going well and what is not going well – and this kind of small assessment can often be tackled as part of the release retrospective. In this case, the assessment is a tool that helps the team assess themselves and potentially measure improvement progress.

Enterprise assessments, on the other hand, can take months.

Large, Fortune 500, projects with an IT application portfolio consisting of hundreds of applications and initiatives can take several months with several assessors, conducting hundreds of face-to-face interviews with various project participants and stakeholders, several workshops and facilitation sessions, blind surveys, review of hundreds of project artifacts, endless data analysis, and finally a C-suite level presentation of the findings.

The point here is that Agile assessments can vary widely in scope and purpose.

Some companies simply might want to know how they are doing with their Agile / Scrum transformation as compared with a baseline. Others might try to use an assessment as a tool to identify targets for training budgets. And others might want to identify non-performing projects.

The focus and purpose varies.

As pointed out earlier, this kind of assessment tool is easily put together. The key to its usefulness is how you conduct the assessment.

When conducting assessments special care has to be taken so project participants do not feel threatened. Especially with a third party coming into a project team or assessing multiple teams, confidentiality is critical so that everybody can be assured that their honest feedback will not be used against them.

When dealing with a large group of project participants and stakeholders across multiple teams, anonymous surveys can be used to validate findings.

Last but not least, transparency is key. Findings should be shared within the team(s), and conclusions and recommendations should be made public. This is critical so that future assessments will receive the same level of cooperation and openness.

Chapter 19
The Persistent Myth of Multitasking and Why Multitasking Does Not Work

Many job applicants proudly proclaim to be "excellent at multitasking." Interviewers look applicants in the eye and ask with a straight face "How good are your multitasking skills?" as if to gauge if you unlocked the secret sauce to productivity. Your boss might bore you with multitasking tales of years past when he/she "worked three jobs at the same time, while studying for my PhD, and...."

Multitasking is commonly acknowledge to be a sign of a highly productive individual or organization. All the multitasking tales are supposed to impress. All the multitasking fanatics are convinced that they are highly productive. And nothing could be further from the truth.

The truth is that multitasking does not work. *It does not work.*

Here is a simple way to verify this yourself:

1. Check your watch and time the following:
    - On a piece of paper, write down "*This is a particularly difficult sentence to spell out backwards!*" – This sentence has 65 characters, including spaces and punctuation marks.
    - Then write down 65 numbers in sequence, starting with 1, going up to 65; like this… 1, 2, 3, 4, 5, 6, … 65
    - You just sequentially finished two tasks… one, writing down the sentence, and, two, writing down the number sequence. How long did it take? – It took the author about 1 minute and 54 seconds.

2. Next, we will try some multitasking! We will take the first task, writing the sentence, and multitask with writing down the number sequence, switching back and forth. Check your watch and time the following:
    - On a piece of paper, write down "T, 1, h, 2, i, 3, s, 4," ", 5, i, 6, s, 7…." and so on, until you completed the

sentence and the number sequence by alternating back and forth (multitasking!)

- How long did it take? – It took the author about 4 minutes and 41 seconds. About 2.4 times slower.

Try it yourself! What are your times?

That's a big difference in time spent. If we now think about adding a third task, like drawing alternating squares and triangles in between, or even a fourth task (whatever it might be), it becomes clear that the total time spent multitasking will most likely be substantially more than if we were to work through each task sequentially.

Now, considering our little exercise above, some will argue that switching back and forth on every letter and every number does not really represent what we experience in the workplace. We are not that interruption driven! Maybe, but regardless if you switch on every letter and every number, or just on every 5th, task switching back and forth will slow you down.

The problem here is the work interruption and overhead imposed with switching back and forth. Each time we interrupt the thought process of one task to refocus on another task, we impose a cost on productivity. The more cerebral the activity, the harder it is to switch back, or, put another way, the higher the cost we impose in terms of time wasted.

Software engineers, for example, need some amount of time – depending on personal ability to refocus, complexity of task, familiarity with the domain, etc. – to get back into the groove of things once they have switched from working on a piece of code. The more complex the code, the harder it is to context switch back and forth. Some software engineers might dispute this, but there is a clear correlation between work interruptions and code quality. The more interruptions, the buggier the code. This is worth reflecting upon, as in this situation we are not just producing things slower, but we are producing things of lesser quality, introducing downstream problems. That is the proverbial double whammy. In this case multitasking not only takes more time but introduces additional problems down the road.

Based on a study performed at UC Irvine, about 82 percent of all interrupted work is resumed on the same day. But here's the bad news — the research showed that it takes an average of 23 minutes and 15 seconds to get back to the task[35].

So, multitasking does not work. Multitasking is one form of work interruption, but any kind of interruption is bad for productivity. This includes emails, pop-up notifications, meeting reminders, phone calls, people walking into the office to ask you something, loud work environments that distract you one way or the other, etc. The key thing is to avoid being interrupted.

There are very few of us who actually can multitask. Before you go off into a celebratory dance and high five all the other multitaskers out there, be aware that we are talking only about approximately 2% of people[36]. For us remaining 98% out there, we are out of luck!

Considering the workplace realities we all face today, considering

---

[35] https://www.ics.uci.edu/~gmark/chi08-mark.pdf

[36] http://www.businessinsider.com/multitasker-test-tells-you-if-you-are-one-of-the-2-2014-5

the endless opportunities for all kinds of interruptions we live with today, what are we to do? Fact is that you will not be able to avoid answering your boss's phone call or text message, you will not be able to ignore your baby crying while you work from home, and you will not be able to deny your colleague who asks for help.

You will have interruptions that you cannot control. That is a given. Too many things are not under your direct control for you to shape the environment in the ideal way.

However, the good news is that if you limit work interruptions and organize your work sequentially, and create an environment where quiet work time is available at least part of the day, you have a very good chance to actually improve your productivity significantly.

Here are some commonsense things that have worked for the author:

1. Block "work time" on your calendar.
2. During your work time, turn off all notification-capable devices and applications.
    - Turn on Do Not Disturb on your mobile phone.
    - Turn off any other notification that might pop up on your computer.
3. Turn off / mute your desk phone (if you still have one).
4. Exit your email client and any application that is not necessary to work through the work task at hand.
5. Close your office door or wear headphones if you sit in a cube.
6. Whatever enables you to create some uninterrupted work time

to boost your productivity does the trick. Just do not call it multitasking any longer!

CHAPTER 20
PRODUCT OWNERS – WHAT MAKES A GOOD ONE?

Much has been written about what makes a good Agile / Scrum product owner. There seem to be endless discussion threads dealing with just this subject.

The following table shows a cross-section of highlights randomly pulled off the internet when googling "*what makes a good product owner*"; basically they show what are considered "must have" personality traits or characteristics of a good product owner:

Key Characteristic	Role	Action
Represents the business	Customer Delighter	Clearly expresses the connection between larger business goals and small backlog items
Has a vision for the product	Story Teller	Write user stories in a way that allows the team to contribute to the "what"
Makes responsible decisions	Delegator	Modify the backlog in response to change without having to discard lots of work
Provides visibility to business leadership	Knowledge Broker	
Maximizes business value	Effective Escalator	
Motivates technical teams to perform to their fullest potential	Conflict Resolver	
Engages through leadership	Visionary and Doer	Set priorities
Available within reason	Team Player	Trust the team
Informed about the product	Communicator and Negotiator	Own mistakes for himself and team; give credit freely
Empowered through humility	Leader	Lead by example
Agile in all things		
Fun and reasonable		

Table 15 - What makes a good Product Owner?

Although we could spend days analyzing each of them, things are much simpler than that.

1. Product owners need **business acumen** – that is a given. Product owners need to be domain experts or at least understand enough of the domain to pull information effectively from others. If the product owner does not understand the business, he really cannot own the product!
2. Product owners need to **understand the basic Agile / Scrum process** they are operating within – that is a given also. Although many organizations make detailed knowledge of Agile / Scrum a must-have requirement when hiring product owners, it is fair to say that anybody interested can acquire Agile / Scrum process knowledge in two days of study or by attending a CSM / CPO class.
3. Product owners need excellent **people and communication skills** – decide, communicate, facilitate, arbitrate – these are at the core of what a product owner needs to do.

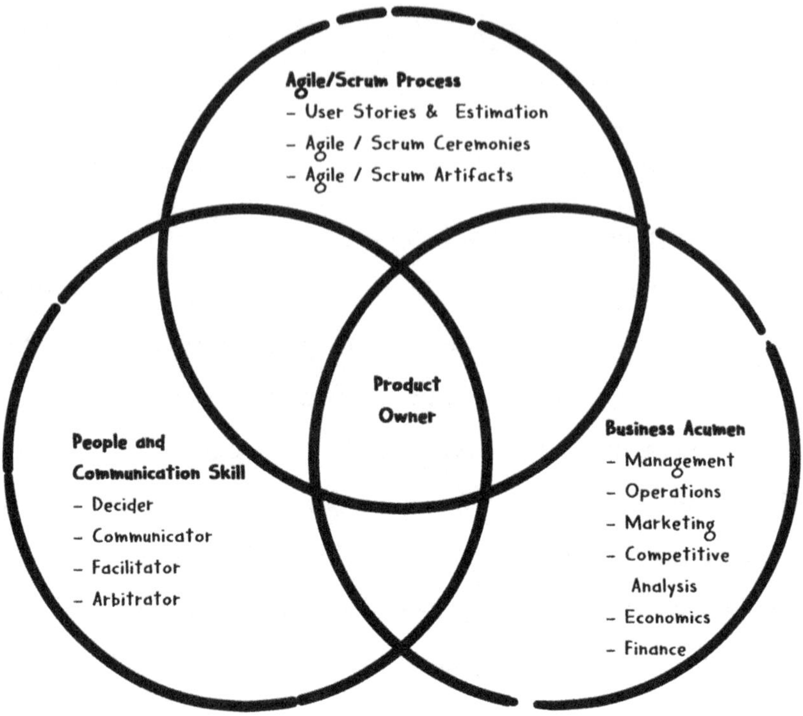

Figure 81 - Product Owner Skills

In summary, do not overthink the role of product owner. Do not load up your job description or position requirement with superfluous items that do not directly add value to the product owner role.

Product owners provide strategy and direction for the project based on their business knowledge. This direction becomes ever more granular top-down, from product vision, product roadmap, and release goal, to sprint goal. At the lowest level, the product owner organizes and prioritizes the product and sprint backlogs on an ongoing basis.

Product owners initiate requirements based on changing market conditions, perform scope control, and act as the main interface between the Agile / Scrum team and the business regarding requirements and status.

Product owners decide on the release date for completed functionality and usually are responsible for the profitability of the product (ROI), ultimately making investment and financial trade-off decisions.

They do participate in daily sprint activities to provide clarifications, which basically means they are a full-time member – acting as the product owner usually cannot be a part-time job.

It is recommended to not get too much into the weeds on any of the more esoteric issues, for example, the level of granularity product owners need to reach when documenting user stories.

The reason for this is simple: They pale in comparison to the two main factors that a product owner needs to bring to the table to be successful:

1. Business acumen
2. People and communication skills

Find the right product owner with **business acumen** and strong **people and communication skills**, and all the other issues will work themselves out.

Chapter 21
Scrum Masters - What Makes a Good One?

Scrum masters are famously described as "Servant-Leaders." Top-notch, effective, scrum masters are hard to find.

Here is a common punch list of scrum master duties and skill sets:

- Is a servant leader
- Responsible for monitoring and tracking
- Responsible for reporting and communication
- Is the process master / facilitates the scrum process
- Is the quality master

- Resolves impediments / helps resolve impediments (works with others to resolve)
- Resolves conflicts
- Shields the team from external interference
- Provides performance feedback to the team
- Creates an environment conducive to team self-organization
- Captures empirical data to adjust forecasts / velocity measurement
- Enforces time boxes
- Keeps scrum artifacts visible
- Promotes improved engineering best practices
- Has no management authority over the team
- Has a leadership role

Effective scrum masters know Agile / Scrum. They must have a sense of responsibility and "ownership" for their team and the team results without having the authority to enforce anything. They usually have excellent communication and teamwork skills, allowing them to work effectively across the development team, the scrum team, the project team, and the organization's influencing constituents.

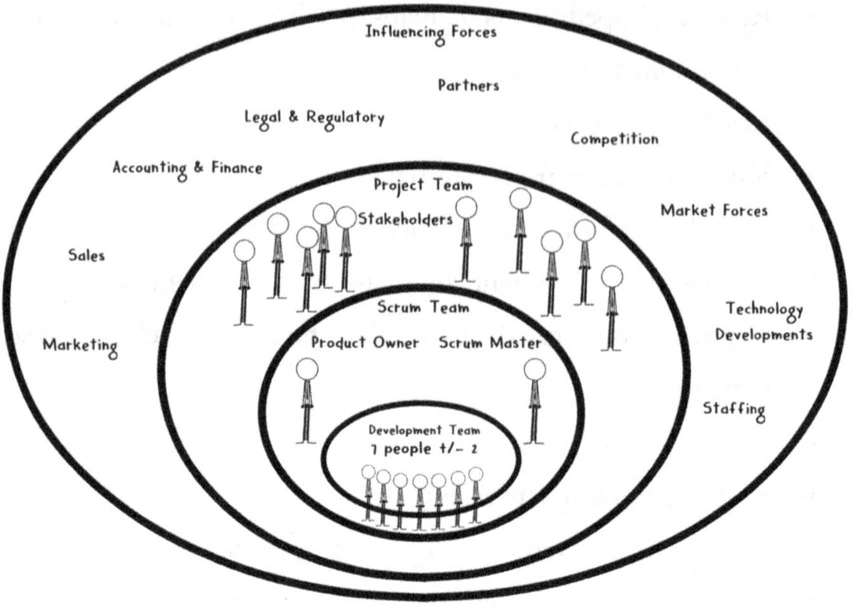

Figure 82 - Teams and Influencing Forces

In order to move their team as well as the organization along, they need persistence and a high level of determination. This goes hand in hand with being a change agent, both in relationship to their team and any project / product stakeholders they need to influence. Many Agile / Scrum transition efforts have failed because the scrum masters involved did not have the right skill set in this area.

Scrum masters must be patient enough to help make the changes happen one at a time, as change is usually a series of incremental improvements. Positive process improvements take time, and scrum masters need both the patience to deliver them as well as the influencing skills to explain why it takes so long.

Figure 83 - Scrum Master as Buddhist Monk

Scrum masters need to be able to challenge others and be comfortable being challenged by others; their ability to ask for help, especially from higher up in order to remove obstacles, is useful but often a difficult task requiring political finesse.

Scrum masters must be able to hold to their process convictions (the Agile / Scrum process), their own skills, and their team's ability to deliver. They are supposed to be the steady rock amongst the waves of uncertainties that development activities sometimes pose.

Scrum masters require a unique personality profile to be a "servant-leader."

They need to have selfless, monk-like, dedication to the team and accomplishing its goals, watch out for the team, remove impediments, and frequently navigate politically difficult waters.

Figure 84 - Scrum Master as Enforcer

Figure 85 - Scrum Master as Director

All this while having the strong leadership traits like General Patton.

So, this job requires a unique skill set, and therein lies the problem – on the one side, the original intent and qualification profile for scrum masters are challenging, while on the other hand, the market is being flooded with certified scrum masters with very little actual knowledge or the skills required.

Acquiring basic Agile / Scrum knowledge is a matter of a couple of days of self-study. Certified scrum master courses are notoriously easy to pass, allowing you to get your CSM certification in about 2 days.

Anybody can pass the CSM certification, but how effective will your scrum master be? There are lots of question that need to be answered:

- Does the scrum master have the right personality?
- Does the scrum master have the right technical skill sets?
- Does he / she know Agile / Scrum?
- Can he / she practically influence the team to delivery within an effective Agile / Scrum process?
- Can the scrum master influence stakeholders to support the team?
- Is he / she able to remove impediments?
- Has the right soft skills to work collaboratively?
- Has enough of a hard edge so as to not appear as a pushover?
- Has specific technology and business skills relevant to the Scrum / Agile project under development?
- Has experience with other SDLC methodologies and can draw on prior professional experience?

SCRUM MASTER VS. PRODUCT OWNER

Product owners really only need to have three key personality traits:

1. A product owner needs to be business savvy and essentially be a domain expert.
2. A product owner needs to understand the basics of the Agile / Scrum process.
3. Good product owners are decisive; they make decisions quickly in order to enable the team to move along with the implementation.

There are excellent product owners who were not particularly good from an Agile / Scrum perspective, and there are great product owners who were quite difficult to deal with in terms of personality. Whatever their shortcomings, good product owners know their business and make decisions. If they know what they want and can communicate their wishes, it works.

Scrum masters, on the other hand, require a complex mix of soft skills, experience, and personality traits that are hard to nail down and even harder to hire for.

Some scrum masters look perfect from a resume / CV perspective, but turn out to be complete ineffective because they do not have the right set of skills or their background does not fit the current challenges they were facing.

Do not forget that the environment a scrum master operates in matters a lot. A particular scrum master can be effective in one environment (say, a fast moving iOS application development project in a startup) but completely ineffective in the next (say, a Fortune 100 bank working on a legacy replacement project). Be aware of such nuances.

HIRING TIPS

When considering someone for a scrum master role, follow this punch list to help make a determination:

- Does the individual have the ability to focus the team on the tasks at hand (sprint focus) while considering the big picture (release focus)?
- What barriers has the candidate removed for past teams?
- What are at least two examples of the "servant" role that he / she filled in the past? For example, has the scrum master helped out the product owner by writing user stories, or prioritizing the product backlog? Tested the product?
- What are at least two examples of the "leader" role that he / she filled in the past? For example, has the scrum master encouraged adopting engineering best practices such as continuous integration, automated unit testing, or pair programming?
- What process improvement suggestions failed? Why?
- What is a good example of the potential candidate influencing stakeholders? How? What were the results?
- How does the scrum master deal with criticism?
- What are some examples of the scrum master candidate

coaching a prior team? How did he / she coach, what was the subject, and what was the outcome?

- Does the candidate exhibit intellectual curiosity? Is he / she smart?

Agile / Scrum emphasizes people over process, as is evident in the focus of a scrum master.

Agile / Scrum emphasizes delivering customer value over extensive documentation and other non-value-added artifacts and processes. That is reflected in the scrum master's emphasis on engineering best practices and focus on delivering working software.

Agile / Scrum promotes open communication and active contributions from team members and the product owner throughout the project. An effective scrum master accomplishes this by encouraging and facilitating ongoing verbal collaboration, both formally and informally.

Scrum masters are human, too, and as such will make mistakes – and anybody who makes mistakes will have to be able to deal with criticism gracefully. As the saying goes, if you cannot stand the heat, get out of the kitchen. Thin-skinned scrum masters do not make effective servant-leaders.

Last but not least, if you find a good scrum master, hold on to him / her! This is a difficult role to fill, and keeping a good, effective, scrum master around will go a long way to ensuring your teams will succeed utilizing the Agile / Scrum process in the most productive way.

CHAPTER 22

AGILE DEVELOPMENT TEAMS - WHAT MAKES A GOOD ONE?

This chapter is about how to build and maintain a good Agile / Scrum development team. It is technology agnostic out of necessity. Agile / Scrum as a process framework is used across the entire IT / software technology space and across all kinds of industries, so we cannot address specific technologies or skill sets required based on certain architectures. Feel free to modify, extend, or enhance based on your specific circumstances.

Figure 86 - Happy Teams are Productive Teams

SOFT VS. HARD TECHNICAL SKILLS

Technical skills are changing at an ever increasing rate. Technical skills are trainable and transferable.

Certain core technical skills, such as database administration, have become homogenized to the point where similarities outweigh differences (say between Oracle, MySQL, and SQLServer). Also, many of the core programming languages now support similar feature sets, to the point where knowing one language enables you to pick up another one with only minimal retraining.

Building and maintaining an effective development team is not a matter of interviewing team members for their particular technical skill sets such as Java, Objective-C, C#, Oracle, or SQLServer – what matters most is that team members are

- Personable
- Experienced
- Reliable

- Communicative
- Cooperative
- Freely share
- Solve problems
- Committed
- Respectful
- Supportive
- Fun to be around

In short, team members need the right **soft skills** to be effective. For example, if you are an effective scrum master utilizing Jira, you also most likely will be effective as a scrum master utilizing VersionOne or Microsoft TFS. If you are a top notch database administrator on Oracle 11, you most likely will be effective on Microsoft SQLServer 2014. As such, it makes little sense to focus the interview questions predominantly on technical skills.

Just to be clear, technical skills are important, but they are secondary to other skills and characteristics that are essential in order to be an effective team member.

Also, specific technical skills are more important for junior level engineers, as it allows them to get started and become productive on their new job, but senior level engineers often have no challenge switching databases, SQL dialects, programming languages, or OS platforms.

HOW TO INTERVIEW TECHNICAL TEAM MEMBERS

Technical qualification questions should not make up the majority of an interview. You should focus no more than one-third of the candidate selection criteria on actual detailed technical questions.

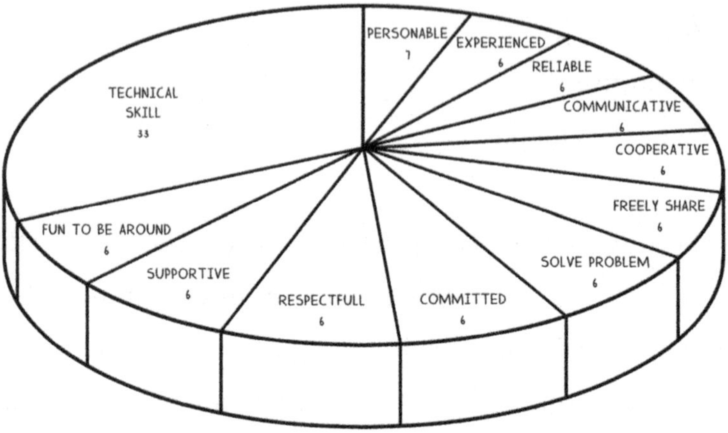

Figure 87 - Ideal Developer Skills Distribution

Technical qualifications can often be asserted within a one-hour group interview process. Group interviews are most effective because all participants see the candidate in the same circumstances and hear the same answers. This is a much better approach than the usual interview round-robin process where a candidate goes from one interviewer to the next and answers the same questions over and over again.

Group interviews are also ideal to see how the potential candidate interacts with the group of future peers.

Technical knowhow can often be effectively validated during an interview with a simple checklist process – assuming the interviewers are technically capable.

What is important is that the interview should not stop there! Look beyond the technical qualifications. Try to gauge the potential of the candidate. Some things to look out for during a group interview process that help determine the soft skills are questions like the following:

INTERPERSONAL SKILLS

- Describe how you developed relationships with others when you were new on your current/ most recent job.
- Have you ever worked for an extremely talkative manager? How did you ensure you were communicating effectively?
- Have you ever worked for a very reticent manager? How did you ensure you were communicating effectively?
- Describe a time when you had problems with a supervisor and had to communicate your unhappy feelings or difficult disagreements. Tell me what you did and what happened.
- What words do your current co-workers use to describe you, and how accurate do you think these are? If they are not accurate, what do you think caused the person to choose that word?
- Tell me about a time when you became involved in a problem faced by a co-worker – how did it happen, what did you do, and what happened?
- When you are dealing with co-workers or customers, what re-

ally tries your patience and how do you deal with that?

- Describe a situation when one of your decisions was challenged by higher management; what did you do and how did you react?
- Describe how you changed the opinion of someone who seemed to have a very negative opinion of you.
- Describe some of the most unusual people you have known – what was different about them and how did you work with them?
- Give me an example of how you handled a very tense situation at work.
- How skillful do you think you are at "sizing up" others? Give me an example.

ORAL COMMUNICATIONS

- What types of experiences have you had dealing with irate customers or clients?
- What was the hardest "sell" of a new idea or method you have had to make to get it accepted?
- Tell me about a time when you "put your foot in your mouth" and what happened.
- Describe a problem person you have had to deal with at work; what did you say?
- What has been your experience in dealing with the poor performance of subordinates?

- All of us feel shy or socially uncomfortable at times – when have you felt this way about communicating? Has that influenced your career?
- Timing is very important in communications sometimes; tell me about a time when your timing was very good and one when it was bad.
- The old proverb "silence is golden" is sometimes hard to live by – tell me about a time when you were proud of your ability to restrain yourself from saying something.

TEAMWORK

- Tell me when you think it is important for a manager to use a participative style and involve work unit members in making decisions.
- Describe the types of teams you have worked in and tell me what worked well and what did not.
- Have you ever had to work with a team of people who did not work well together or did not like each other? Tell me what happened and how you reacted.
- Give me an example of a situation in which you managed or led a team and were able to create a high morale, high productivity work group.
- Tell me about a time when you had difficulty getting others to work together on a critical problem and how you handled it.
- Some managers encourage people to work together while oth-

ers encourage direct competition among their people. Tell me about the best manager you have worked for and what approach that person took and what you learned.

- Describe a really difficult person you worked with and how you handled working with that person.

PROBLEM-SOLVING

- Describe a time when you were able to reverse a very negative situation at work.
- What are the critical factors you look for in evaluating the work of others? How did you develop these concepts and how, specifically, do you use them?
- Describe a major work problem you faced and how you dealt with it.
- Describe a situation in which you found yourself to be very good at analyzing a problem and deciding on a course of action.
- Tell me about a time when you solved a problem through intuition and how that happened.
- Give me an example of a problem you have faced and how you solved it.
- Tell me about a time you tried to solve interpersonal problems of co-workers.
- Describe a problem you faced that was almost overwhelming and how you got through it and kept from being completely

overwhelmed.

- Describe a situation in which you had to solve a problem without having all the information you needed – what did you do and what happened?

One final note about soft skills and personality types. Type A and Type B personalities contrast each other.

According to Wikipedia,[37] personalities that are more competitive, outgoing, ambitious, impatient and / or aggressive are labeled Type A, while more relaxed personalities are labeled Type B.

Type A personalities often are perceived as rude, ambitious, rigidly organized, highly status-conscious, sensitive, impatient, anxious, proactive, and concerned with time management. People with Type A personalities are often high-achieving "workaholics." They push themselves with deadlines, and hate both delays and ambivalence.

Type B individual traits contrast with those of the Type A personality. Type B personalities, by definition, are noted to live at lower stress levels. They typically work steadily, and may enjoy achievement, although they have a greater tendency to disregard physical or mental stress when they do not achieve. When faced with competition, they may focus less on winning or losing than their Type A counterparts, and more on enjoying the game regardless of winning or losing. Unlike the Type A personality's rhythm of multi-tasked careers, Type B individuals are sometimes attracted to careers of creativity: writer,

---

[37] https://en.wikipedia.org/wiki/Type_A_and_Type_B_personality_theory

counselor, therapist, actor or actress. However, network and computer systems managers, professors, and judges are more likely to be Type B individuals as well. Their personal character may enjoy exploring ideas and concepts. They are often reflective, both personally and on the team level.

Be careful not to select the majority of your team members who are one or the other. You will need both personality types on your team.

Figure 88 - Too many Type A personalities make for loud discussions

Too many Type A personalities can cause teams to implode in a flurry of arguments and competitiveness. Too many Type B personalities can make the team seem less dynamic and achievement driven. You need to find the right mix for your environment.

Seniority and division of labor also flows into selection process. Once you have determined what skills you need on your team, you need to think about how those skills map to the team members that you are selecting for your Agile development team – in short, what will be your team composition?

TEAM COMPOSITION

Every organization has some kind of career progression approach, where employees enter the company at a junior level and progress up through the ranks. A typical skills / experience pyramid looks something like this:

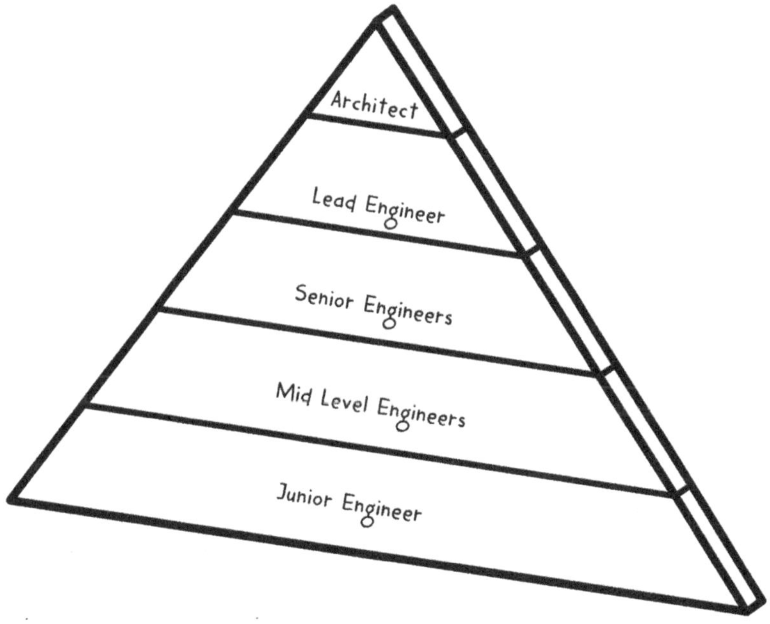

Figure 89 - Development Team Sill / Experience Pyramid

As you may notice, there are no job titles like scrum master, product owner, or development team member mentioned here. Many Agile / Scrum proponents argue that companies should get rid of career ladders like the one above, but the reality is that this is still the predominant way most HR departments deal with career paths.

Until companies align their organization completely along Agile principles, we will be stuck with something like this. If you like to see some additional viewpoints about this, check out the chapter: "***Lack of Organizational Realignment.***"

Once you know your skill sets and the job titles that are in use, the question shifts to how to most optimally assemble a team based on the Agile / Scrum guidance to make the development team size be around 7 (+/-2)?

Here are two examples of how you can put together a team.

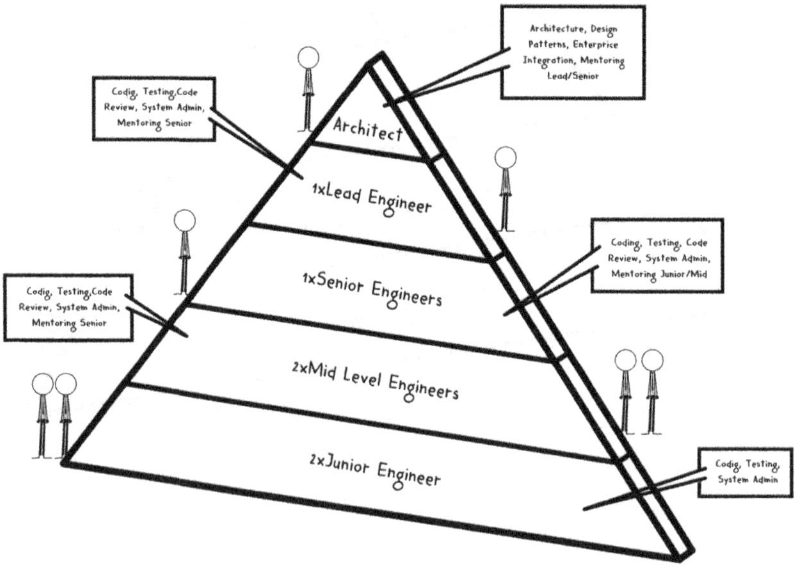

Figure 90 - 2-2-1-1-1 Development Team

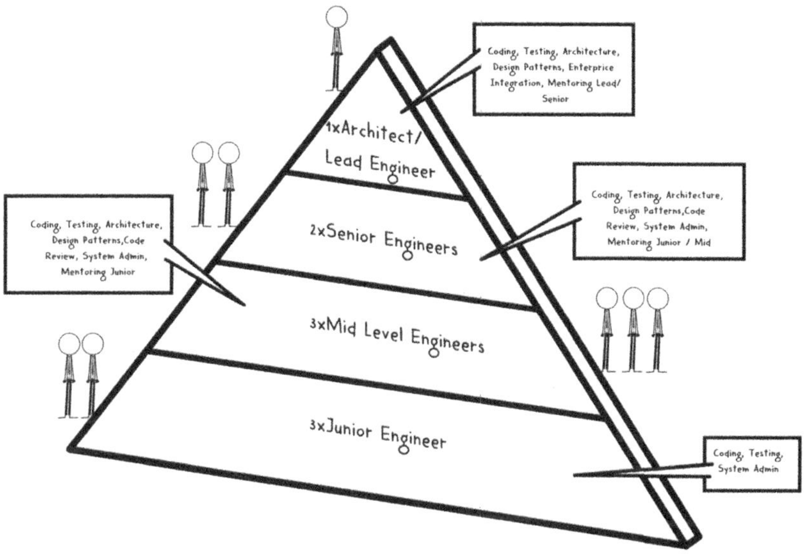

Figure 91 - 3-3-2-1 Development Team

These are just samples. You will need to think through your particular situation in order to see what is feasible.

SINGLE POINTS OF FAILURE (SPOFS), REDUNDANCY, AND CROSS-TRAINING

A big part of establishing, and maintaining, an Agile / Scrum development team is to ensure coverage across various skill sets, if possible with varying degrees of experience.

Single points of failure should be avoided at all cost, for the obvious reasons that people get sick, transfer to another department, and sometimes resign. Whenever one team member is the only one capable of doing a job, it usually is not good sign. Whenever one team member insists on being the only one to do a certain activity, or maintains his / her solitary expertise on a skill set, that is a bad sign as well.

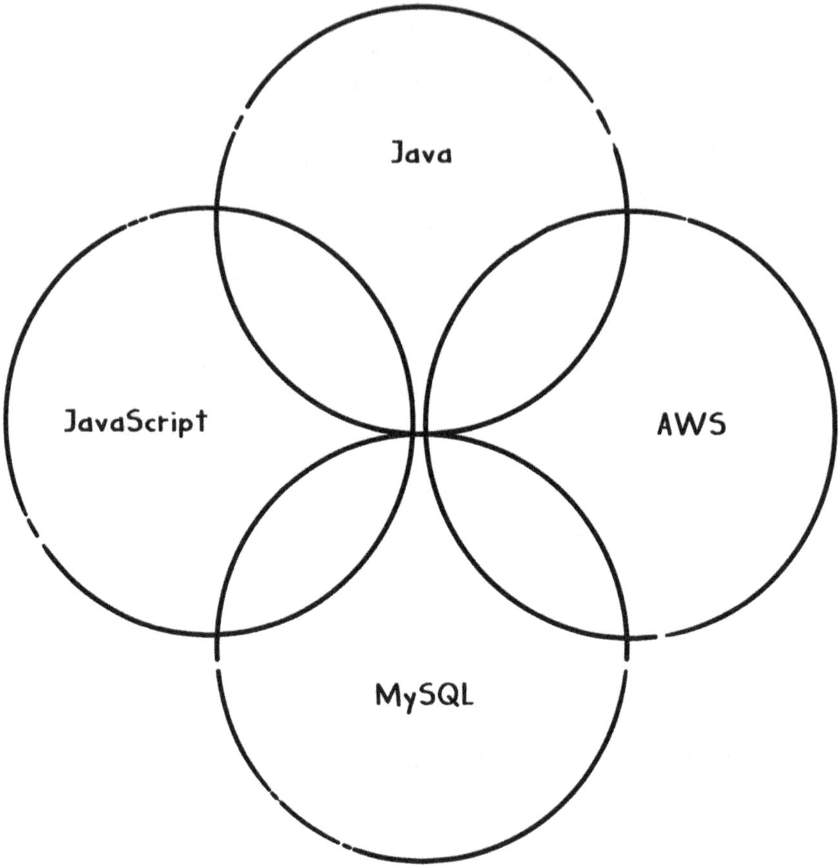

Figure 92 - Sample Skill Set Overlap

Agile / Scrum is all about collaborating, working together, and sharing – including skill sets. Hence, the focus on the "free sharing" soft skill mentioned above.

Cross-training each other is an important characteristic of a successful Agile / Scrum development team because it creates skills redundancies that enable the team to keep moving forward. It does not matter if this is a formal activity, or an informal "lunch and learn" kind of setup – as long as it happens.

The team should never be caught off guard because somebody is absent. Work should never stop because there is a single point of failure.

TEAM COMPOSITION ANTI-PATTERNS

The following illustrations show some common anti-patterns. All team composition anti-patterns come down to variations of one or several of these.

The first anti-pattern addresses "top heaviness," meaning the team has too many senior level leaders but not enough lower level working engineers.

Sometimes one can observe this due to junior team members "aging up," meaning they joined the team as a junior member but after several years of experience have become more mid-level or senior level without their position on the team being adjusted.

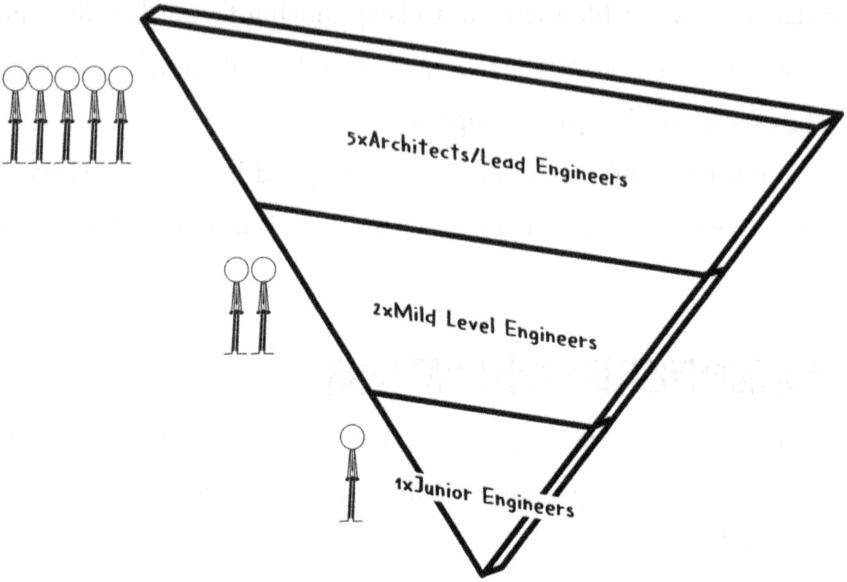

Figure 93 - Top Heavy Anti

Sometimes this pattern can be observed because management is concerned about the product / project, meaning all senior level expertise available is allocated to one project because it is perceived to guarantee success.

Please note that seniority by itself is not the problem; the challenges come in if senior level engineers do not want to perform what is perceived as simpler, grunt level, engineering tasks usually assigned to junior engineers – that is when dysfunction sets in and tasks do not get done. Testing famously is in this category for dysfunctional teams.

As long as senior level folks are willing to work on whatever is necessary to get the job done, this is fine. But in the long run, usually some dissatisfaction sets in. This "top-heavy" model ultimately really is not sustainable.

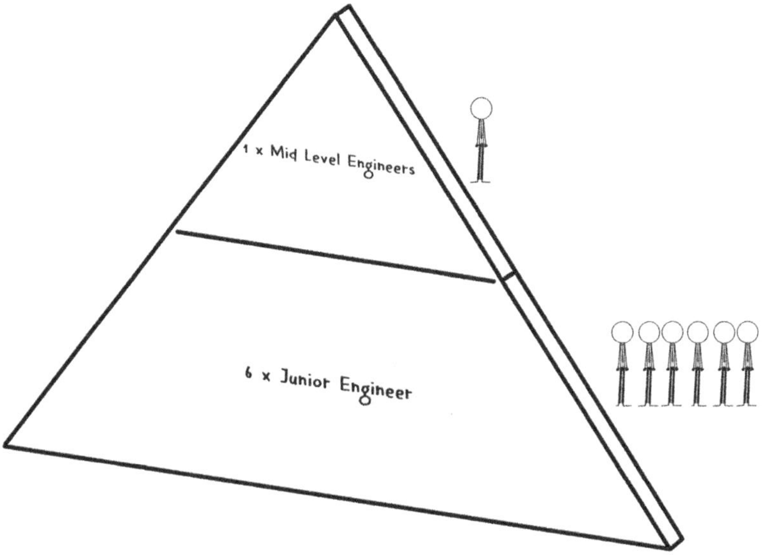

Figure 94 - Too Inexperienced Anti Pattern

The second most common anti-pattern is the reverse of the first; it essentially deals with too many junior level folks on the team who do not get enough guidance, training, and mentoring from more senior team members. The Agile / Scrum team overall is too inexperienced.

This pattern is common with non-technical managers staffing an Agile / Scrum team "to budget," meaning that budget limitations cause the hiring manager to put in too many junior level folks without regard to what is technically necessary.

The third most common anti-pattern deals with extreme geographic distribution of an Agile / Scrum team. There are different kinds of geographic distribution; some are worse than others. The best or the easiest kind to deal with is geographic distribution within one time zone. The worst kind of geographic distribution for an Agile development team to deal with is extreme distribution across many time zones.

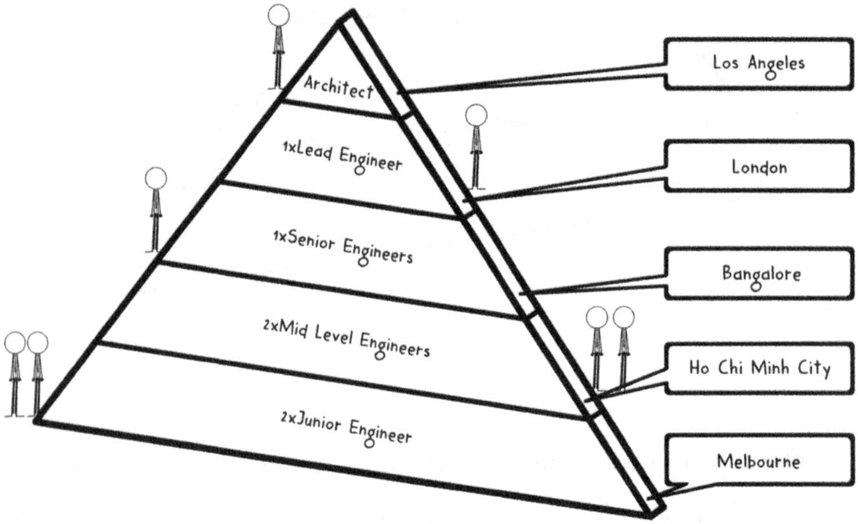

Figure 95 - Extreme Geographic Distribution Anti Pattern

Extreme geographic distribution can have many reasons, such as:

- Skill sets are not available locally.
- Skill sets / domain expertise is only available in other company locations.
- The company is geographically distributed and wants to utilize

available staff members in various locations to build a virtual team.

- Budgets force the hiring manager to build a virtual team with team members in lower cost labor countries such as India or Vietnam (international labor arbitrage).
- Part or all of the development effort is outsourced to partners in different locations.

This anti-pattern is one of the most challenging to deal with. Agile / Scrum originally was intended to enable small, co-located, teams to cut down on communication challenges, quickly determine the right course of action, and rapidly deliver value. Extremely distributed teams suddenly impose a high overhead on communication and decision making – contrary to what Agile / Scrum was intended for.

Making distributed teams successful is not impossible – but it is a lot harder than dealing with co-located teams, and you need to put special focus and emphasis on team structure and communications.

A WORD ABOUT MAINTAINING GOOD TEAMS

Maintaining good, effective, teams over the long run is a tricky process, as you need to balance several different demands.

- On the one hand, you want to maintain team stability – you want as few people as possible to leave, move to another department or take some other responsibility outside of the development team.

- On the other hand, you do want to create some kind career progression and upward, as well as lateral mobility so development team members do not feel "stuck."
- And then you have to deal with competitive pressures, local market conditions, and budget realities.

The truth is that you will need to engage in a constant rebalancing act in order to maintain an effective development team. This rebalancing involves understanding a clear picture of all technical challenges represented by your projects, contrasted by the soft and hard skill required to deliver against those challenges.

This rebalancing might feel like you are never done analyzing, interviewing, and hiring – because you really never are.

This chapter tried to highlight the complexities of building and maintaining an Agile / Scrum development team.

Building a team is a complex undertaking, starting with selecting the right soft and hard technical skills, mapping those skills to your project's requirements, deciding on your ideal team composition, and finally addressing realities such as team distribution and budget considerations.

And just when you think you are done, you will have to consider rebalancing the team, for one reason or the other.

It is a never ending process that will require constant attention, because nothing is more important in Agile / Scrum than your development team!

With an effective development team you can survive a poor

scrum master, you can compensate for a poorly performing product owner, and you can work through all the process challenges.

However, the reverse is **not** true – if you have a poorly performing development team, the best scrum master, running the best process, with a well-qualified product owner will not be able to save the project.

CHAPTER 23

AGILE COACHES - WHAT MAKES A GOOD ONE?

On almost every Agile / Scrum transformation effort, Agile coaches seem to be somewhat elusive, hard to define, team members. There is a lot of confusion about what Agile coaches actually do, and how they benefit the Scrum effort. There are many anti-patterns frequently seen regarding this role, from actually using them as project managers, through viewing Agile coaches as scrum masters running several teams, to utilizing Agile coaches simply as Agile / Scrum trainers.

Following are some explanations that hopefully will make things clear.

Besides the core Agile / Scrum development team, there are three major roles that need to be filled for an Agile transformation effort:

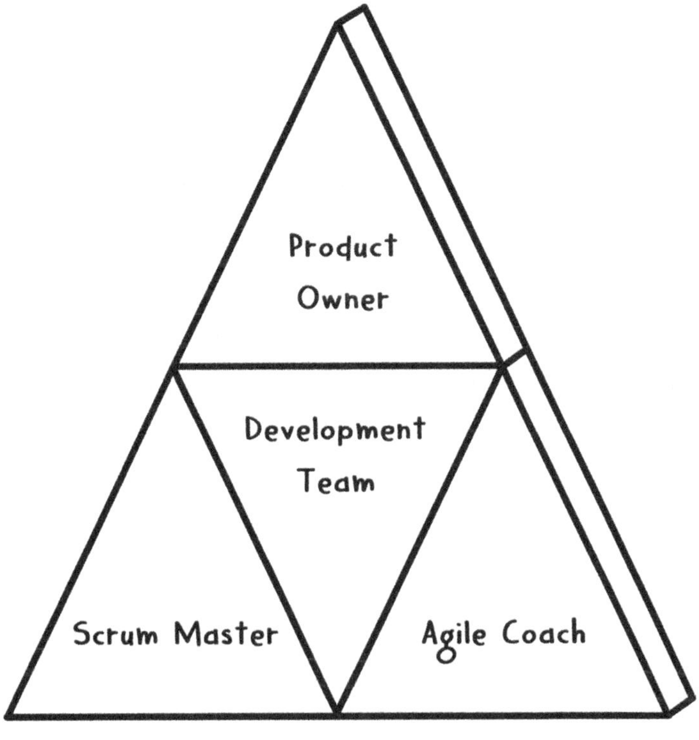

Figure 96 - Agile / Scrum Roles

In essence, the roles are defined as follows:

- The **scrum master** is the owner of the process and quality, the "servant-leader" focusing on making his / her team as productive as possible on a day-to-day basis.

- The **product owner** is the domain expert, with the intricate business knowledge to define both the big picture (vision and roadmap) as well as the product details (user stories).
- The **Agile coach** is focused on working with customers / project teams in a creative process that ***inspires their professional potential***; in order to unleash that full potential. Agile coaches frequently facilitate amongst groups that help them come to solutions and make decisions.

So, what does this mean for an Agile coach in an Agile / Scrum context? The following illustration shows core Agile / Scrum competencies.

AGILE / SCRUM COMPETENCIES

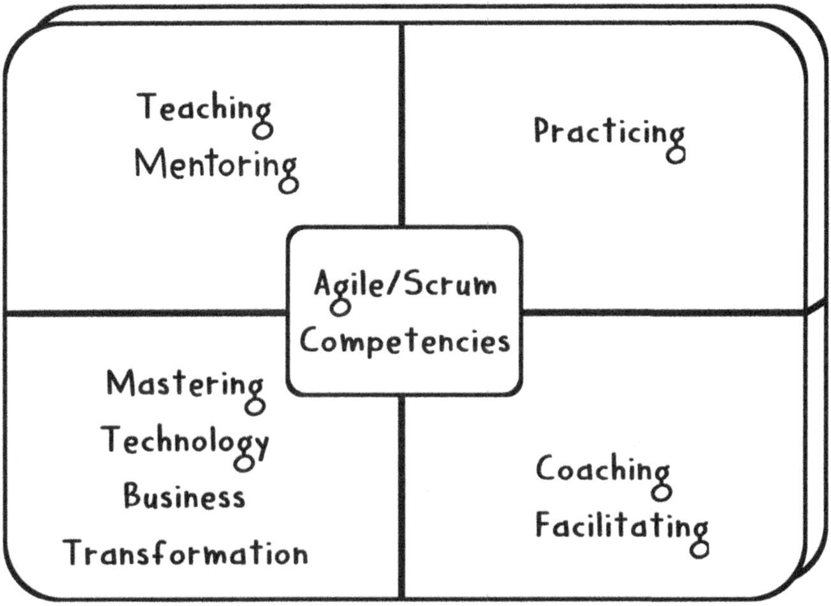

Figure 97 - Core Agile / Scrum Competencies

Agile coaches operate in the **coaching / facilitating** quadrant. Let us quickly review the Agile / Scrum competencies and how they can be defined:

- **Teaching** – represents the ability to offer the right knowledge, at the right time, taught in the right way, so that individuals, teams, and organizations metabolize the knowledge for their best benefit.
- **Mentoring** – represents the ability to impart one's experience, knowledge. and guidance to help grow another in the same or similar knowledge domains.

- **Practicing** – represents the ability to learn and deeply understand Agile frameworks and Lean principles, not only at the level of practices, but also at the level of the principles and values that underlie the practices enabling appropriate application as well as innovation.
- **Mastering Technology** – represents the ability to get your hands dirty architecting, designing, coding, test engineering, or performing some other technical practice, with a focus on promoting the technical craft through example and teaching-by-doing.
- **Mastering Business** – represents the ability to apply business strategy and management frameworks to employ Agile as a competitive business advantage such as Lean Start-Up, product innovation techniques, flow-based business process management approaches, and other techniques that relate to innovating in the business domain.
- **Mastering Transformation** – represents the ability to facilitate, catalyze, and (as appropriate) lead organizational change and transformation. This area draws on change management, organization culture, organization development, systems thinking, and other behavioral sciences.
- **Coaching** – represents the ability to act as a coach, ***with the customer's interest determining the direction***, rather than the coach's expertise or opinion.
- **Facilitating** – represents the ability to guide the individual's, team's, or organization's process of discovery, holding to their

purpose and definition of success.

It is worth pointing out two key phrases used to highlight what an Agile coach actually does:

1. **Inspires the professional potential in his / her customers and project teams.** To use a sports analogy, if you run the 100-meter dash in 22 seconds, the coach's job is to inspire you through various techniques to get you to run faster than that – according to your athletic potential. It is not his / her job to make you break the current world record of 9.58 seconds.
2. **Coaches with the customer's interest determining the direction.** Every business and every project is different. It is not up to an Agile coach to determine how the business should be conducted – as such the coach needs to take guidance from the customer and coach accordingly.

The reason to point this out is simple. Too many times Agile coaches come into organizations and quickly proclaim that this or that is wrong without fully understanding the customer's interests or direction, nor really wanting to bring out the greatest potential in teams. Those two areas are at the core of what an Agile coach does.

Also, it is incumbent upon the Agile coach to extract from the customer exactly what the customer's interests and intent are – sometimes this can be a feat in and of itself. A good coach helps customers distill and articulate this information.

If you asked a therapist, a consultant, and a coach "How do I learn to ride a bike?"

The therapist would have you approach the bike and ask what experiences have you had with riding bikes.

The consultant would give you instructions to follow.

That coach would say, "OK, get on the bike and I will support you until you can do it on your own."

One final note about competencies. Facilitation is an important skill / technique that good Agile coaches bring to the table. However, an excellent facilitator might be a horrible Agile coach. Agile coaches need to be grounded in Agile / Scrum, technology, and business, in order to successfully navigate the coaching process.

The same goes for training. You might have a really good scrum trainer who presents the Agile / Scrum subject matter concisely and effectively, with great humor. But this does not mean that the trainer is an effective Agile coach (or scrum master or product owner).

HOW TO FIND A GOOD AGILE COACH

When looking for a good Agile coach, here are some of the things to look out for.

1. **Charismatic personality** – coaches do not all need to have the same stage presence as Tony Robbins, but they do need to have a presence and the confidence that goes along with it to influence people across the organization, from team members to executives.

2. **Approachable** – while a good coach needs to have charisma and presence, they also need to be approachable, meaning they need to be equally effective in front of audiences and executives as well as communicate well with staff on a one-on-one basis.
3. **Communicative** – communication, facilitation, and arbitration are all core competencies of an Agile coach.
4. **Experienced** – good coaches have relevant professional experiences; transformation coaches especially should have been through several process transformations before; otherwise you are really hiring somebody to learn on the job. Prior experience in related functional areas is a plus (for example, having worked as a software developer, project manager, scrum master, etc.).
5. **Broad knowledge** – deep understanding of their craft is a must. An Agile coach who is helping a customer improve upon their process not only needs to understand Agile / Scrum, but also needs to have a deep understanding of common software development methodologies (Waterfall, Spiral, Iterative, V-Model, RUP, SAFe®, TDD, XP, etc.), the most current management / business approaches, and be at least conceptually up-to-date on relevant technology developments. Look for someone with practical experience and knowledge acquired in the trenches – not book knowledge.
6. **Available toolbox** – Agile coaches who are worth their money have a set of tools they can pull out of their hat because they

have been there, done that. They can jump into ad hoc working sessions, conduct training, or point you to articles they wrote about common challenges encountered during their prior coaching engagements. Anybody telling you that they will "put something together in a couple of weeks" means they have not been doing the job.

7. **Fit** – think about how the Agile coach fits into your organization. Does he / she have the right experience, industry background, and knowledge to effectively participate and influence?
8. **Curiosity** – does your Agile coach ask you questions? Does he / she understand what drives your desire to bring in an Agile coach? Is he / she fully committed to your vision and guidance when coaching?

PART III

AGILE ADOPTION PITFALLS

The following chapter is the first of several chapters that show common Agile Adoption Pitfalls and how to avoid them.

Over the last decade Agile and Scrum development frameworks have emerged as the de facto development methodologies. This is true not just within the software development industry but in many diverse industries, from biotech to manufacturing.

Whereas 10 to 15 years ago applying an Agile or Scrum process in a disciplined way would have been a great advantage – maybe even given companies a competitive edge – today it is a necessity. If you are not following Agile / Scrum – and you have to follow it correctly in order to reap the benefits – you are falling behind. It's as simple as that.

The core business drivers that are pushing this adoption cycle are the traditional **better – faster – cheaper** motivators. And because many of your competitors are doing things better, faster, and / or cheaper, you really do not have an option but to go all out and adopt Agile / Scrum.

Companies proudly proclaim that "We are doing Agile!" However, upon review, these companies fall short of truly applying an Agile / Scrum process in an effective way. Or, they applied Agile / Scrum but did not seem to realize any productivity gains. So, the obvious questions are "Why not?", "What are the most common problems companies face?", and "How can we avoid those common pitfalls?"

Those are the questions that the following chapters will try to answer.

Chapter 24
Methodology Fatigue

Not to quote Donald Rumsfeld, but too many companies never ever think about the "unknown unknowns"; they assume that Agile / Scrum is simply another development methodology, another fad, just like the others before it.

After all, anybody who has done this for a while knows that after Waterfall came V-Models and iterative methodologies, then the Rational Unified Process, various frameworks that testified to solve the world's problem. So why should Agile / Scrum be different?

Here is a quick refresher of the most commonly used development methodologies.

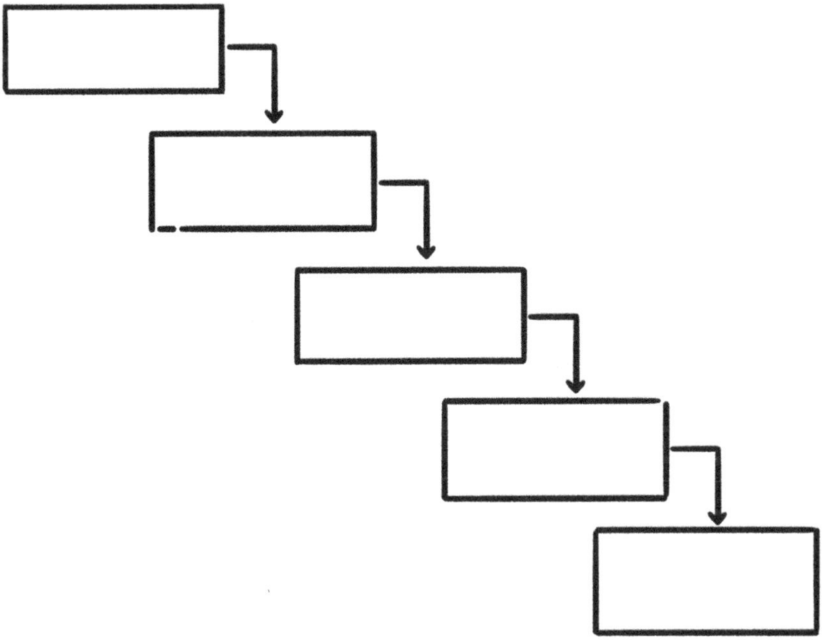

Figure 98 - The Waterfall

WATERFALL

The **waterfall model** is a sequential (non-iterative) design process, used in software development, in which progress is seen as flowing steadily downwards (like a waterfall) through the phases of conception, initiation, analysis, design, construction, testing, production / implementation, and maintenance.

The waterfall development model originates in the manufacturing and construction industries: highly structured physical environments in which after-the-fact changes are prohibitively costly, if not impossible. Because it was created in a time when no formal software development methodologies existed, this hardware-oriented model was simply adapted for software development.[38]

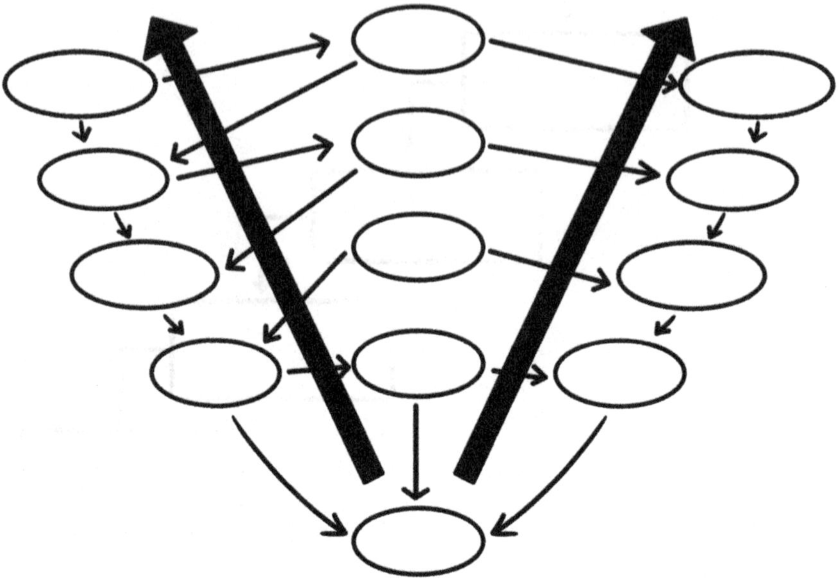

Figure 99 - The V-Model

V-MODEL

In software development, the **V-Model** represents a development process that may be considered an extension of the waterfall model. Instead of moving down in a linear way, the process steps are bent upwards after the coding phase, to form the typical V shape. The V-Model demonstrates the relationships between each phase of the development life cycle and its associated phase of testing. The horizontal and vertical axes represent time or project completeness (left-to-right)

---

[38] Benington, Herbert D. (1 October 1983). "Production of Large Computer Programs" "Production of Large Computer Programs" (PDF). IEEE Annals of the History of Computing. IEEE Educational Activities Department

and level of abstraction (coarsest-grain abstraction uppermost), respectively.

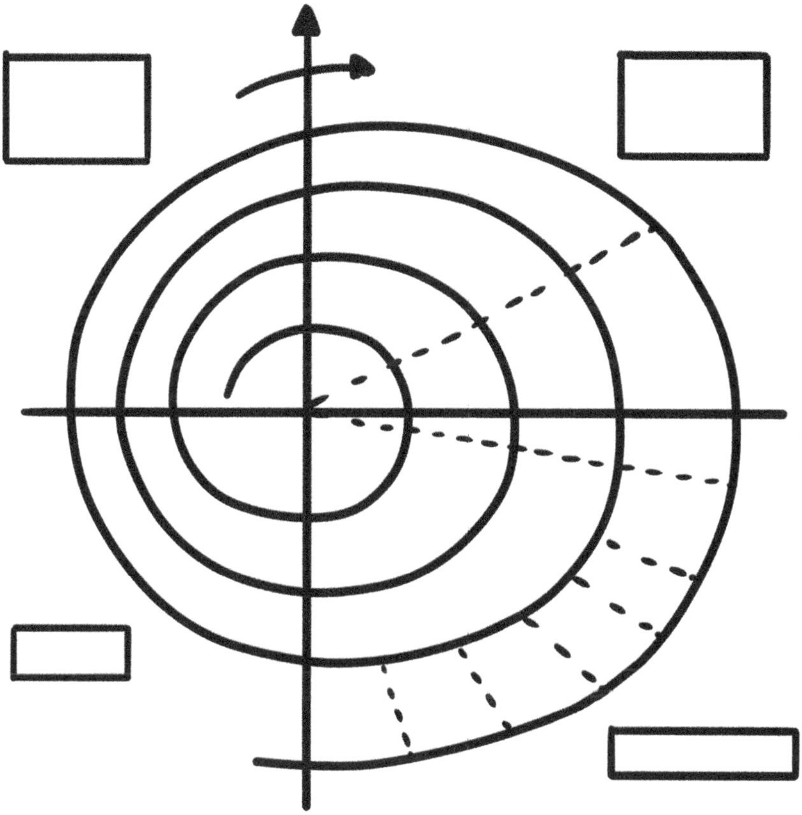

Figure 100 - The Spiral

SPIRAL

The **spiral** model is a risk-driven process model generator for software projects. Based on the unique risk patterns of a given project, the spiral model guides a team to adopt elements of one or more process models, such as incremental, waterfall, or evolutionary prototyping.

This model was first described by Barry Boehm in his 1986 paper "A Spiral Model of Software Development and Enhancement"[39]. In 1988, Boehm published a similar paper to a wider audience[40]. These papers introduce a diagram that has been reproduced in many subsequent publications discussing the spiral model.

These early papers use the term "process model" to refer to the spiral model as well as to incremental, waterfall, prototyping, and other approaches. However, the spiral model's characteristic risk-driven blending of other process models' features is already present then.

[39] Boehm B, "A Spiral Model of Software Development and Enhancement Boehm B, "A Spiral Model of Software Development and Enhancement", ACM SIGSOFT Software Engineering Notes, ACM, 11(4):14-24, August 1986

[40] Boehm B, "A Spiral Model of Software Development and Enhancement Boehm B, "A Spiral Model of Software Development and Enhancement", IEEE Computer, IEEE, 21(5):61-72, May 1988

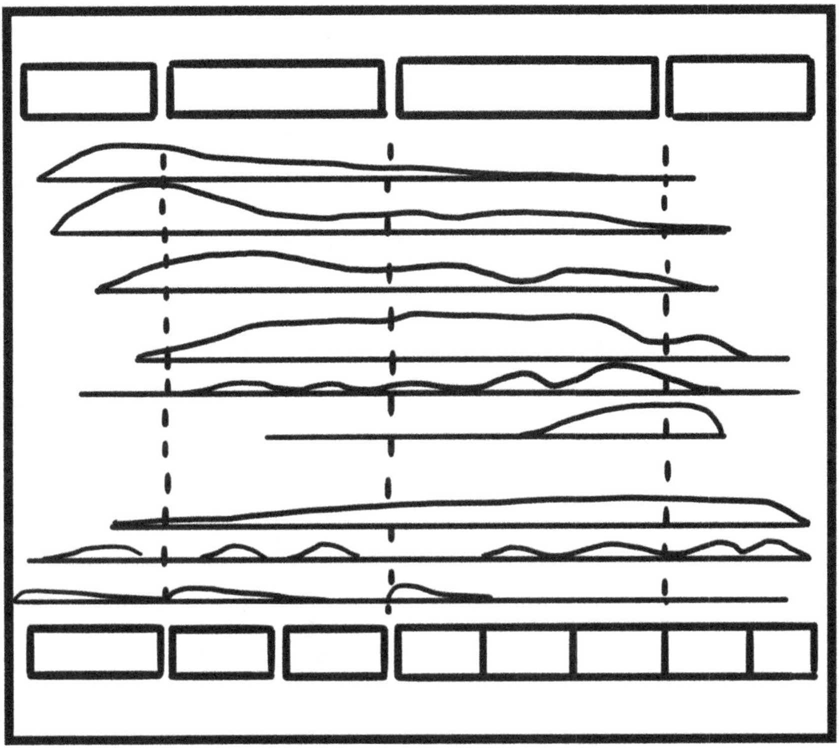

Figure 101 - The IBM Rational Unified Process (RUP)

IBM RATIONAL UNIFIED PROCESS

The **Rational Unified Process (RUP)** is an iterative software development process framework created by the Rational Software Corporation, a division of IBM since 2003. RUP is not a single concrete prescriptive process, but rather an adaptable process framework, intended to be tailored by the development organizations and software project teams that will select the elements of the process that are appropriate for their needs. RUP is a specific implementation of the Unified Process.

The RUP determined a project life cycle consisting of four phases (Inception, Elaboration, Construction, Transition). These phases allow the process to be presented at a high level in a similar way to how a waterfall-styled project might be presented, although in essence the key to the process lies in the iterations of development that occur within all of the phases. Also, each phase has one key objective and milestone at the end that denotes the objective being accomplished. The visualization of RUP phases and disciplines over time is referred to as the RUP hump chart.

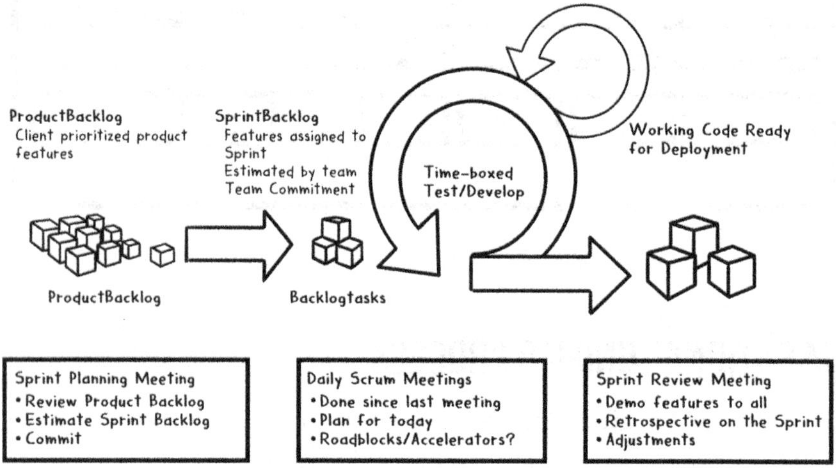

Figure 102 - Agile / Scrum

AGILE

Agile software development describes a set of principles for software development under which requirements and solutions evolve through the collaborative effort of self-organizing cross-functional teams[41]. It advocates adaptive planning, evolutionary development, early delivery, and continuous improvement, and it encourages rapid and flexible response to change[42]. These principles support the definition and continuing evolution of many software development methods[43].

The term Agile was adopted by the authors of the Manifesto for Agile Software Development (often referred to as the Agile Manifesto for short)[44].

As mention above, iterative and incremental software development methods can be traced back many decades. Evolutionary project management and adaptive software development emerged in the early 1970s.

During the 1990s, a number of lightweight software development methods evolved in reaction to the prevailing heavyweight methods that critics described as heavily regulated, planned, and micro-managed. A parallel technical development in software engineering that facilitated this discussion was the rapid adoption of PC technologies and the general adoption of object-oriented programming paradigms.

Lightweight development frameworks or methodologies includ-

---

[41] Collier, Ken W. (2011). *Agile Analytics: A Value-Driven Approach to Business Intelligence and Data Warehousing*. Pearson Education. pp. 121 ff. ISBN 9780321669544. What is a self-organizing team?

[42] "What is Agile Software Development?" Agile Alliance. 8 June 2013

[43] Larman, Craig (2004). Agile and Iterative Development: A Manager's Guide. Addison-Wesley. p. 27. ISBN 978-0-13-111155-4

[44] Kent Beck, James Grenning, Robert C. Martin, Mike Beedle, Jim Highsmith, Steve Mellor, Arie van Bennekum, Andrew Hunt, Ken Schwaber, Alistair Cockburn, Ron Jeffries, Jeff Sutherland, Ward Cunningham, Jon Kern, Dave Thomas, Martin Fowler, Brian Marick (2001). "Manifesto for Agile Software Development". Agile Alliance

ed: from 1991, rapid application development (RAD); from 1994, the unified process and dynamic systems development method (DSDM); from 1995, Scrum; from 1996, Crystal Clear and extreme programming (XP); and from 1997, feature-driven development. Although these originated before the publication of the Manifesto for Agile Software Development, they are collectively referred to as Agile software development methods[45].

And these are just the big flavors! One could easily list 50 variations, about 15 for Agile alone (XP, SAFe®, Kanban, TDD, etc.), which basically resulted in "methodology fatigue" across many industries.

Also, be aware that each old methodology that was replaced by the most current one left behind a tail of methodology artifacts, sometimes referred to as "methodology junk." Just like junk in the real world, these remnants are no longer useful, but they linger on. Some examples include UML diagrams (specific to RUP), arcane ticketing systems that have waterfall built into the process, requirements documentation templates that do not match the commonly used Agile / Scrum user story format, or management reports that reference metrics no longer relevant.

The point is that oftentimes Agile / Scrum is adopted as the new methodology but it is expected that old systems and processes remain in place – often resulting in Agile / Scrum being forced into the old process framework – limiting the level of success that a new Agile /

---

[45] Larman, Craig (2004). *Agile and Iterative Development: A Manager's Guide.* Addison-Wesley. p. 27. ISBN 978-0-13-111155-4

Scrum process could actually deliver on. This is a typical scenario and usually ends with somebody commenting sooner or later that "...after all this Agile stuff, nothing changed; we still can't get things done faster...." Which should not really be a surprise.

Consequently, when companies start Agile / Scrum efforts they often do not think through the challenges ("... yeah, it's one of those new methodologies...") and jump into the adoption with:

- **Lack of buy-in at the right levels** – Agile / Scrum adoption only works if it is sponsored and supported from the executive level, top-down, **and** from the line level, bottom-up. The reason for this is simply that an extensive process and culture change represented by an Agile / Scrum adoption cannot be achieved by only forcing it top-down or by trying to stage a revolution bottom-up. Everybody has to buy into the process for it to effectively improve productivity.
- **Lack of professional transition support** – "Start doing Agile!" is a common battle cry trying to fast track the adoption, instead of starting an Agile / Scrum adoption process with a good pilot project and the assistance of knowledgeable Agile coaches; start small and scale up virally by demonstrating success.
- **Lack of training** – Both at the executive / managerial as well as at the engineering / line level, many companies do not train their employees about what Agile / Scrum actually means. Let me be clear: everybody needs to be trained on Agile / Scrum in

order to make it work, and to guarantee that expectations are aligned; rinse and repeat training sessions frequently, with new hires and current employees, in order to solidify their understanding. You know you have done a good job training the organization when the CEO, the Director of HR, and the line level Tech Support Engineer can talk about sprints, user stories, and retrospectives.

- **Lack of process review and adjustment** – basically simply plugging in Agile / Scrum as just another methodology without reviewing and adjusting supporting systems or processes (such as management reports) will not work. Thoughtfully reviewing and streamlining the current process in a way to optimally utilize the benefits that Agile / Scrum requires a dedicated effort.
- **Lack of support for learning** – Agile / Scrum assumes that teams "self-calibrate" and continuously learn. As such, Agile / Scrum teams sometimes fail, and failure is a big part of learning. If the company does not support learning this way, Agile / Scrum efforts are doomed. Frequently, instead of embracing failure as a way to learn, companies immediately start criticizing the new Agile / Scrum approach as the most likely culprit.
- **Unrealistic expectations** – executives, managers, and line level engineers often go into an Agile / Scrum adoption process with completely unrealistic expectations – all with different unrealistic expectations, though! Executives expect magical productivity gains in short order; managers often assume that

Agile / Scrum will not impact their functional responsibilities (nothing could be further from the truth); and engineers frequently hope that Agile / Scrum will mean the end to tedious analysis and documentation tasks – often a direct reaction to the previous methodology not having worked so well. Setting these expectations and creating a common understanding is imperative to success.

Chapter 25
Lack of Organizational Realignment

The previous chapter covered **methodology fatigue**; this chapter will deal with the **lack of organizational realignment**. What does that mean? Why do you have to realign the organization to become more Agile? And what do you have to realign?

There are really four main areas that require serious cross-functional and cross-departmental review when jumping into an Agile adoption process. These four areas represent core company processes, assets, and values that need to be adjusted for realignment if the Agile adoption process is to be successful.

The reason we refer to them as core company values or established processes is because many companies, especially established ones, follow a set of practices without necessarily understanding why – the historical perspective and original reasons for doing things one way or the other have been lost. These are areas that impact many aspects of the business or the relationship between a company and its employees. Agile adoption oftentimes exposes all those challenges, so rethinking these areas is critical.

So, what are the four main areas? They are:

1. **Human resources**
2. **Facilities**
3. **Communications technology**
4. **Product & business strategy**

Let us review each in detail.

HUMAN RESOURCES

Human resources – especially job descriptions, career paths, and career development – completely ignores Agile / Scrum adoption considerations. HR usually ignores an Agile / Scrum adoption initiative (to their defense, most times nobody tells them!), while the Engineering / Software Development organizations happily proceed to focus internally, as if there are no other considerations. But there are many HR-related considerations that, if not taken care of, seriously impact the long-term viability of any Agile / Scrum adoption process. For example:

Reviewing functional organizations, their job descriptions, and career paths, you will find that many are based on older Waterfall / V-Model / Iterative models; you will most likely see job titles such as:

- Technical Support Analyst I, II, III
- Quality Assurance Engineer I, II, III
- Product Manager
- Program Manager
- Sr. Program Manager
- Software Engineer I, II, III
- Software Architect
- Software Project Lead
- Team Lead
- Test Lead

Once you go into an Agile / Scrum adoption initiative, you will see job titles like this

1. Scrum Master
2. Product Owner
3. Developer
4. Agile Coach

So, the core question is how to transition traditional jobs, such as Software Architect or Project Manager, to new Agile / Scrum job descriptions. The answer is: *very carefully* – you need to design the new job descriptions with great care considering the mode of operation that Agile / Scrum introduces. There is no standard list of job descriptions that will fit all companies, big and small. However, there are common themes to consider:

1. In order to redesign job descriptions, make sure to train HR on Agile / Scrum principles. Without basic knowledge of Agile / Scrum, HR will be lost on how to go about this.
2. In Agile / Scrum, the team takes precedence over individual contributions. Job descriptions have to reflect the fact that some team member will primarily focus on making the team more productive instead of focusing on individual "quarterly / yearly accomplishments."
3. Career progressions from junior to senior to principal still apply, including training and cross-promotion opportunities.
4. Review current employees for their "fit" and assign them new roles. If people do not fit the new roles / job descriptions, lay out training plans to get them up to speed. In some cases you might not be able to transition people into key roles, so you might need to consider hiring new people with the right skill set.
5. Last but not least, you can backfill open spots temporarily with knowledgeable consultants, especially in roles focused on Agile / Scrum transition and training.

The result of all this job description redesign is that you will most likely create "pods" of scrum masters, "pods" of team members, "pods" of product owners, and "pods" of specialists. Beware that in order to make this transition, you need to explain the logic and intention behind all this – meaning you need to explain, explain, and explain again. Do not assume that everybody is happy with the transition to Agile / Scrum, especially outside of core engineering functions.

There are some job transitions that come more naturally that others. All of them require individuals to refocus from a "document-centric" world view, where things are documented and passed on to the team downstream, to a "communication-centric" world view, where folks work closely with each other to discover what needs to be built and resolve problems immediately. Here are some high level fits that have been observed:

1. **Program managers** make good **scrum masters** because they are used to running interference across multiple departments and are good at moving the teams along toward delivery.
2. **Project managers** often do not feel comfortable managing an Agile / Scrum project because they miss their tight-knit Phase Gate Approval Process and their Gantt Charts. This is not to be facetious here, just to point out that the author has observed many traditional project managers struggle with the Agile / Scrum approach. This is especially difficult for organizations that have large, established PMO groups, with heavy reliance on PMI certifications. Agile / Scrum upends much of

the project management gospel that is preached.

3. **Software team leads** usually can play various Agile / Scrum roles, capable of acting as both **scrum masters** and **product owners.**
4. **Software architects** can act as **product owners.**
5. **Product managers** can act as **product owners** if they feel comfortable with the real-time elaboration process; product managers used to writing extensive Requirements Specifications Documents often struggle with the tight and straightforward user story format or the sprint-by-sprint refinement process.

Please do not take these observations verbatim. You need to assess who fits what role for your specific environment. However, it is strongly suggested that organization stay away from "Wagile" (Wagile = Waterfall + Agile). Too many times one can come across job titles such as:

- Project Manager / Program Manager / Scrum Master
- Agile Project Manager
- Scrum Master Manager
- Program Agile Leader
- Product Marketing Manager / Product Owner
- Director of Engineering / Scrum Master
- Etc.

The above job titles are surefire signs your organization is on the wrong path in your Agile / Scrum adoption process! You are basically pretending that you can fold the new Agile / Scrum process into whatever you already have, with minimal changes, and achieve stellar productivity gains. Unfortunately, it's not that simple.

Also, once you have redesigned the titles and job descriptions, do not forget to rethink and reassess your incentive programs – old models do not work any longer. How do you incentivize "team orientation"? Again, there is no magic list here that you can simply follow; your HR team and your management team need to put their thinking hats on to come up with creative solutions. In terms of incentive programs, it is recommend to tie them to "business value created," which is at the core of Agile / Scrum.

FACILITIES

Facilities – reconfiguring work space to match the Agile / Scrum process. This sounds simple but can be very challenging.

When we talk about facilities, what do we actually mean? Basically how teams and people sit and work together, considering the optimal office layout, conference rooms, break rooms, rest areas – in general the best appropriate physical environment. Sad truth is that many companies miss the boat on this one. One reason is that tearing down existing office space and reconfiguring it is not that simple.

The balance you need to find is between creating an environment that allows for open communication and flow of ideas while avoiding the trading floor style pandemonium that distracts software engineers from creating good code. Optimizing communication while maintaining a quiet work environment is the devil in this detail.

One thing that has been observed to work is an environment with cubes, where team members sit in close proximity together while still affording the privacy of your own space. Combine this with lots of conference rooms, and you have the opportunity to create a work space that provides for open communication.

Conference rooms need the latest communication equipment, video conferencing technology, and lots of white boards. Conference rooms need to range in terms of their usefulness from full-blown auditoriums to dedicated team meeting rooms to "phone booths" where people can step out to take a call.

Also helpful is a mobile phone policy that stresses that it's unacceptable to take your mom's phone call next to the engineer trying to code up the latest device driver. Take the call outside; or even better, the company should provide enough conference rooms of abovementioned "phone booth" size.

What does not seem to work are "old style" office layouts that are based on functional titles – the bigger the title, the bigger the office. Fact is that oftentimes you do not have enough offices to accommodate everybody, so cubicle setups are more common now. And, you need to plan for reconfiguring space as projects develop; Agile / Scrum office layouts are more project focused, not permanent. Again, it's not uncommon to see decent office layouts that would make Agile / Scrum teams highly productive, but people refuse to move, resulting in team members of a new team being spread over multiple locations.

This is really a management and expectation settings issue – if you work in Agile / Scrum teams, be prepared to move as you progress through projects. If you expect to stay stationary in the same cube for years, Agile / Scrum is not for you.

COMMUNICATIONS TECHNOLOGY

Communications technology and communications approaches are often ignored. It cannot be stressed enough what big differences exist in productivity between companies that go all out with their communication technology vs. the ones that try to save money. Here are some concrete examples:

- **WiFi connections** – if your WiFi is slow, go and invest in the best possible WiFi solution you can get; nowadays, every Agile / Scrum team member can easily have four devices that connect to the WiFi network, so saving here is counterproductive.
- **Internet speeds** – get the fastest possible internet connection.
- **Poor VPNs** – along the lines of poor internet speeds, VPNs frequently end up being too slow to enable effective work from home or while traveling. As with WiFi and internet speeds, companies need to bite the bullet and invest in the fastest VPN technology / configuration available.

In terms of communication approaches, one must train all Agile / Scrum team members to communicate in order of effectiveness (from 1, best, to 5, worst):

1. **Face-to-face** – nothing beats face-to-face communication.
2. **Video conferencing** – if you can't meet face-to-face, video conferencing comes in second, but only if you turn on the camera! The face and body communicate most of the information in a conversation, so if you cannot meet face-to-face, video is the preferred second approach.
3. **Phone** – if you cannot see your counterpart, at least hear them live; again, the live voice communicates nuances that are important to observe and understand.
4. **Text** – text messages / chat are preferable to email because you still have a live connection to your counterpart, although without a visual or audible way of discerning if your message is received or how your counterpart's message was intended.
5. **Email** – the least effective way of communicating is written email or snail mail. You lose all immediacy. And email communication frequently turns a 5-minute discussion into a 7-day email thread.

PRODUCT AND BUSINESS STRATEGY

You jumped into Agile / Scrum, you worked with HR to redefine responsibilities and job descriptions, you tore down walls and reconfigured your work space; but did you think through all the necessary changes in your **product and business strategy**?

Every business is different, but you really need to consider the following areas in order to marry the Agile / Scrum benefits with the business:

1. **Upgrade cycles**, from yearly releases to on-demand, monthly, or quarterly software updates due to cloud technology, continuous integration, faster cycle times – how does that change your business?
2. **Constant support needs** – with faster release cycles, support models change.
3. **Revenue generation** – old, traditional models relied on different revenue generation models; how does a yearly upgrade model change to a subscription model?
4. **Cloud technology** – away from IT to hosted services; how does the cloud help you scale? Do most of your servers sit idle simply to deal with the Thanksgiving or Christmas business spike two days of the year? Do you save money by using the cloud or spend more?

Agility changes everything![46] Thinking that Agile adoption was going to be easy is one of the biggest misconceptions of all!

In conclusion, this chapter tried to show that there are many factors that influence and, in most cases, determine the successful outcome of an Agile adoption initiative – factors that have nothing to do with Agile / Scrum, the projects, the technologies used, the teams, or

[46] https://hbr.org/2013/05/why-the-lean-start-up-changes-everything

the process. But those factors have everything to do with how your organization accommodates the changes that are necessary to be successful.

Chapter 26
Agile Burnout Death Spiral

In many ways this pitfall is the easiest one to spot, and the hardest one to stop – because of the momentum that builds up with it.

The **Agile burnout death spiral** is akin, and closely related, to what Ed Yourdon described in his classic, and still highly relevant book, "***Death March***"[47].

Also relevant to this discussion is another classic by Frederick P. Brooks Jr., "***The Mythical Man Month: Essays on Software Engineering***"[48].

These two books are timeless classics that describe common

---

47 http://www.amazon.com/Death-March-2nd-Edward-Yourdon/dp/013143635X/

48 http://www.amazon.com/Mythical-Man-Month-Software-Engineering-Anniversary/dp/0201835959/

management dysfunctions when doing software projects, and to a large extent explain why software projects are so costly and have such high cost and schedule overruns.

So, what is a common Agile burnout death spiral scenario? Here is an approximation:

1. A project gets initiated – "We need to do XYZ; otherwise we <will lose market share><won't be able to raise more money><will lose credibility with the business stakeholders><etc.>."
2. Usually, right along with the project kickoff, somebody will announce a completion date that is completely devoid of any basis in reality. At this stage, scope has not been clearly defined; nor has the team been assembled to do any estimation of what is involved – "XYZ needs to be done by September 15th."
3. A team is hastily assembled. A part-time project manager is dedicated as the scrum master, and the project start date is today. Because team members aren't available, a mix of permanent employees that can be pulled off other projects, contractors, and third-party consultants are assembled. This results in a geographically dispersed team across five different international time zones (LA, NY, Slovakia, Belarus, India).
4. As there is no knowledgeable product owner available, the VP of Marketing will act as the product owner whenever he has time.

5. Planning, user story definition, and estimation efforts start off chaotically, due to a lack of experienced team members who have worked with each other before and a lack of clear guidance from the product owner, who is barely available to provide clarifications.
6. Time zone differences are the cause for major communication challenges, some of which the team members try to overcome by having late night / early morning conference calls. But coordination efforts are difficult, resulting in many questions and answers flowing through email, which causes major delays and in some cases misunderstandings.
7. Because of the preset end date (September 15th) the scrum master "backs" the sprint planning into the end date and loads up the sprints with user stories regardless of team capacity or realistic estimates.
8. Because sprint planning was driven by the end date and not by estimates, sprints come up short right away – basically the team is not able to deliver at the velocity necessary to achieve the sprint goals.
9. After missing several sprint goals, the scrum master discusses the possibility of slipping the end date. "It doesn't look like we are going to make September 15th."
10. Management panics and gives the go ahead to add additional resource and do "whatever it takes" to make the target date.
11. The scrum team grows to 29 people.
12. Instead of starting to deliver faster, velocity actually decreases,

and confusion increases.

13. The death spiral spins out of control.

After reading this, you must be asking "Why would anybody in the world do this kind of stuff?" The above set of bullet points probably have at least 20 cardinal sins that every scrum master and software executive should be able to list. So why do things like this happen?

Well, it's a matter of perspective. Yes, sometimes management just does not know how to adopt to Agile / Scrum, and you can see dysfunctional behavior like the ones outlined above simply because people do not know better.

However, management sometimes is completely aware of the bad practices that are used; they often understand perfectly well that one, two, three teams will be burned out, people will be lost to resignations, nervous breakdowns, and burnout. **But the business end of achieving a certain goal at whatever cost *might* justify the means.**

There are always multiple sides to a story. Software engineers want to design the best software, in a decent work environment, on the latest technology stack, with all the relevant buzzwords to build up their resumes for the next job.

Business stakeholders, on the other hand, often are not concerned at all how the solutions they are delivering are built as long as they are available on time.

Business executives usually are date sensitive, quarterly driven, paranoid about the competition, driven to get things out to market. Software engineers, on the other hand, are often completely devoid of any concerns about generating revenue or being profitable.

Ultimately, the business question is this: "Is the instability that an Agile burnout death spiral creates in terms of employee turnover and burnout worth the risk in terms of potential future business?" For example, will this project give us an edge over our competition, position us in the market that makes us unique?

What about the software engineers who are stuck in a death spiral project? Engineers will have to ask themselves if the experience they are gaining in terms of technical know-how, other skills acquired, and financial benefits outweighs the personal sacrifice that they have to make. For example, will this project make my 5,000 stock options worth $300,000?

These kinds of death spiral projects happen all the time, sometimes because of lack of knowledge, but sometimes deliberately, for valid business reasons. Oftentimes death spiral projects are early startup or new technology projects, where companies and teams go through a "forming – storming – norming" process. That is totally normal. Just beware of the companies that use this approach as the standard mode of operations.

Chapter 27
The Fizzle

The previously mentioned Agile burnout death spiral is pretty easy to spot.

Just as that one is easy to grasp and understand, **the fizzle** can be the more frustrating to interpret because the visible symptoms are not that easy to identify and understand.

The fizzle is like a high school football jock who looks all promising, bustling with raw talent, but then flames out on the way through college and never makes it to the NFL – you see all the great performances, you observe all the great plays, you can literally see the promising future, but somehow the long-term success is denied.

Here is a common scenario that shows how the fizzle often plays out:

- Executive management hears a lot about Agile and Scrum and likes the idea of moving the organization into a more lightweight, productive, methodology. It announces its support by declaring "Let's do Agile and get things to market faster!"
- Development managers also like the idea of finally adopting Agile / Scrum, and developers had been arguing for years that the current process-heavy approaches slow things down.
- The PMO organization is ambivalent about the idea, but reluctantly supports it as long as "current PMO processes are still followed."
- Management dedicates a development team lead to take the lead in the new scrum master role and to assemble a team to deliver a pilot project.
- After some discussion, a pilot project is picked. It happens to be a non-controversial project that might turn out be interesting, but nobody within the organization really has a stake in it.
- Some of the best developers across the organization are assigned to the team; not only are the developers all senior level experts in various technology disciplines, but they all also have significant domain knowledge, having all worked at the company for more than 8 years.
- A knowledgeable software architect with deep domain knowledge and years of integration experience with the diverse customer base is dedicated as the product owner.
- The pilot project gets underway in dedicated office space, off

the beaten track, and away from daily interruptions. The work environment and setup is close to ideal, allowing for teamwork and quick decision making.

- Tools are selected, scope is defined, user stories are written, estimation poker is played, and sprints are defined. In short, the pilot is off and running.
- Sprints last two weeks.
- Sprint reviews show good progress.
- Team velocity increases over the first four sprints to steady out around 38 story points; once stabilized, the project predictably delivers each sprint goal as promised, much to the surprise of observers.
- Sprint retrospectives resolve obstacles quickly, increasing team productivity.
- After 5 and 1/2 months, the project is delivered, with exceptional quality, a full feature set, and apparently right on target with market demands.
- During the final executive presentation, one VP of Development states, "This would have taken my team 2 or 3 years to produce; we should aggressively roll this Agile / Scrum thing out to the rest of the organization."
- Following the successful pilot, the company rolls out the "Agile / Scrum thing" to the rest of the organization, but it does not adjust team structures, job descriptions, office setup, conference rooms, telecommunication needs, or other critical supporting factors necessary for Agile / Scrum.

- Whereas the pilot project was staffed locally, allowing for direct face-to-face interaction, the company decides to utilize two outsourcing partners in development center in Bangalore, India, and Saigon, Vietnam.
- Existing processes, like a waterfall-oriented PMO process with excessive documentation and phase gate requirements, stay in place. Agile / Scrum teams are supposed to follow the old and the new processes.
- Agile / Scrum roles like scrum master, product owner, and Agile coach are not clearly defined. Consequently, various employees with various backgrounds end up in roles they are not naturally suited for – as a matter of fact, functional job titles stayed the same, while Agile / Scrum job titles never were made official.
- Two years later, during an executive review, the CEO openly says, "I don't get it; the Agile / Scrum pilot was so successful, and now we seem as stuck in the mud as before. Productivity has not improved at all."

At this point, organizations frequently do one of two things:

1. **The organization reverts back to their previous process** and says "Agile / Scrum just isn't for us" or "Agile / Scrum doesn't work for our industry" or comes up with some other explanation. This is less common, as the organization has to openly admit to a failure.

2. **The organization turns into an Agile / Scrum "zombie"** – they keep using Agile / Scrum terms, tools, and verbiage but basically don't follow Agile / Scrum (daily standup meetings turn into status meetings; retrospectives turn into process improvement discussions; sprints last 8 weeks; scrum masters work on detailed project plans, trying to estimate the task level down to the minute; instead of continuous integration, the teams produce weekly builds; instead of test automation, all testing is performed manually; there is a big test cycle at the end of the project; deployments are done by hand, etc.). This is the more common scenario, as it avoids having a hard discussion about the viability of Agile / Scrum in the organization.

So, what happened? As mentioned above, this one is hard to spot, because with all the excitement of the initial pilot, folks lose sight that one pilot does not a successful Agile adoption effort make.

Here are some, but certainly not all, reasons why Agile / Scrum adoption efforts fizzle out:

1. **You cannot replicate the A-Team** – Do not send in SEAL Team 6 and conclude that it will work for the other 30,000 troops. Perform pilots with middle-of-the-road personnel, not superstars – because superstars are not representative of your average skill sets.
2. **You did not ensure both top-down and bottom-up support**

– Agile / Scrum enterprise adoptions only work if there is top-down and bottom-up support. Executives really need to understand what it means, which means at least two days of training for executive level sponsors. Frequent feedback from an Agile coach to the executive team should give progress indicators along the way.

3. **Repeated team training** - Line-level managers and engineers need to understand the realities as well, so training is essential for them to understand what is necessary to switch to an Agile / Scrum process– do not assume that technical folks understand more than the buzzwords! Often they do not! Rinse and repeat training often until the new Agile / Scrum process becomes second nature.
4. **Engineering best practices still count!** Do not try to move into Agile / Scrum if you are still struggling with basic best practices such as requirements analysis, source code control, automated builds, or automated testing. Too often, organizations fail to realize that engineering best practices are the basis for adopting and taking advantages of all the good Agile / Scrum benefits.
5. **You run into adoption resistance** – Realize that some of your staff will be open to change (originators / change agents), the majority will be neutral to change (pragmatists), and some of your staff will resist change (conservers). The latter folks are the ones who would rather keep buying a rotary phone from the Phone Company – until there are no other options availa-

ble. Accommodate them! Do not ignore this segment of your employees, as they can be vocal. It's your responsibility to at least try to convince them.

6. **Your pilot project was not representative** of all the other projects that you usually have to run – either too simplistic, or too complex. Be realistic. Do perform pilot projects that are small enough to be manageable, big enough to represent complexity, important enough to warrant management attention, but not so important to get folks fired – you do not need a pilot effort driven by fear.
7. **You did not organizationally align** for Agile / Scrum – Setting up a pilot for success is one thing, but how do you organizationally align for 30 teams? What are the capital requirements? How easily can you move people? Do you have buy-in from all the relevant support groups (HR, Facilities, Network Engineering)?
8. **You had unrealistic expectations** – Under-promise and over-deliver; do not set the expectations that Agile / Scrum will solve all your problems!

CHAPTER 28
PROCESS OVERLAY

This chapter will describe a common Agile / Scrum adoption challenge that is referred to as **process overlay**. Other folks might refer to it as "double work Agile" or other terms. Essentially, this pitfall deals with the challenges of *adopting* Agile / Scrum and *adapting* Agile / Scrum into an already existing process framework.

Although the advantages of Agile / Scrum are unquestionable, the author is also rooted in the business realities that companies face. Many companies follow yearly planning and budget cycles (certainly if they are public and need to report quarterly numbers), many companies are not organized to be agile in the Agile / Scrum sense, and many companies are following somewhat traditional, more waterfall-oriented, processes that are not easily replaceable.

So, before everybody gets upset and pulls out the proverbial tomatoes and other vegetables to be thrown, let us revisit some assumptions:

1. You are ***not*** a startup –you are working in an established company with already established processes. If you are a startup and you can design your process from scratch, good for you. Having said that, startups that offer a blank piece of paper from which to start are rare – compared to the majority of IT environments out there.
2. You are ***not*** working in a highly regulated industry like medical devices or software needing FAA or FDA certification, or Sarbanes-Oxley. Nor are you working in highly mission-critical environments, like NASA or some military applications. Software solutions in these problem spaces require regulatory hurdles to be overcome that slow the development activities due to health and safety concerns. However, although your company might produce a medical device that has to be FDA certified, it is very unlikely that ALL your company's IT projects have to follow the same stringent process. Too many times people argue that Agile / Scrum does not work, just because there was one mission / life critical application that had to be developed. This argument does not pass the reasonableness test.
3. You and your company are interested in using Agile / Scrum in order to make your development teams more productive

and to get software / IT solutions out to market faster than what you are doing right now – this is an important point. You do not have to adopt an Agile / Scrum process. If your current process works for you, stick with it. Having said that, the author believes that adopting an Agile / Scrum process in many industries nowadays is a necessity in order to stay competitive. Delivery cycles and time to market windows are faster (i.e., shorter) than ever.

Now that all the disclaimers are out of the way, let us quickly review why Agile / Scrum is such an effective delivery framework. Agile / Scrum is a fairly simple process framework that basically consists of:

Few defined **roles**:

- Scrum master
- Product owner
- Development team member
- Stakeholder
- Agile coach / mentor

Few defined **events / process steps:**

- Sprint planning
- Backlog grooming
- Estimation
- Daily scrum

- Sprint reviews
- Sprint retrospectives

Few defined **artifacts**:

- User stories
- Product backlog
- Sprint backlog
- Shippable product increment

In a nutshell, Agile / Scrum looks like this:

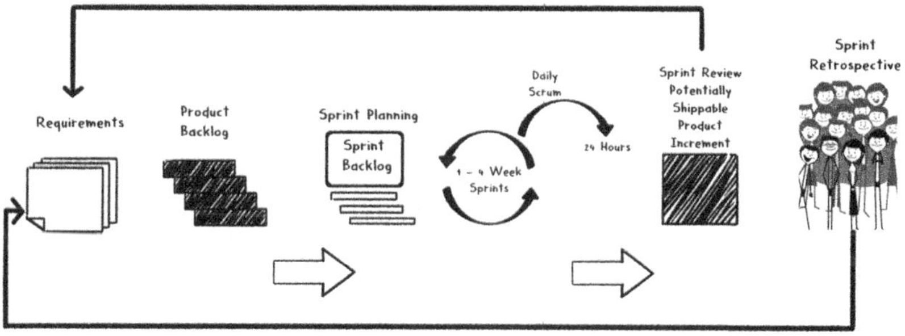

Figure 103 - Agile / Scrum in a Nutshell

The beauty of Agile / Scrum is that it is simple and lightweight by:

- Shortening communication paths
- Empowering teams
- Focusing on working software

- Iterating over a problem space until a good solution is found

It very much follows the ***Agile Manifesto***[49] to the letter of the law and the spirit of the law.

It is understood that the above paragraphs grossly simplify Agile / Scrum, but the author would challenge anybody to succinctly summarize any other methodology / framework in as few words. Agile / Scrum is a very natural and effective process that does not require 255 pages of dense information to grasp.

This chapter tries to address the following problem is: Unless you are blessed with having the opportunity to build a new company / product / team from scratch (see the startup comment above), you are faced with existing processes. This is especially true in medium- to large-sized companies.

Many medium- to large- sized companies follow the well-established PMI PMBOK© project delivery methodology, which roughly defines project phases and supporting processes like this:

1. **Initiating** – Initiation processes help define a new piece of work – either a complete new project or the phase you are about to begin. They ensure you have authority to proceed.
2. **Planning** – Planning processes help define objectives and scope out the work to be done. They also encompass all the work around planning and scheduling tasks.

[49] http://agilemanifesto.org/

3. **Executing** - Executing processes focus on the 'delivery' part of project management, where the main activity happens and you create the products.
4. **Controlling** - Controlling processes track the work that is being done, including the reviewing and reporting on it. They also cover what happens when you find out the project isn't following the agreed-upon plan, so change management falls into this process group. You'll run these processes alongside those in the executing group (mainly, but alongside the other groups too), so you monitor as you go.
5. **Closing** – Closing processes finalize all the tasks in the other groups when you get to the point to close the project or phase.

PMI PMBOK© lays out these five **process groups** and defines **knowledge areas** for each, as follows:

Knowledge Areas	Project Management Process Group				
	Initiating Process Group	Planning Process Group	Executing Process Group	Monitoring and Controlling Process Group	Closing Process Group
4.Project Integration Management	4.1 Develop Project Charter	4.2 Develop Project Management Plan	4.3 Direct and Manage Project Work	4.4 Monitor and Control Project Work 4.5 Perform Integrated Change Control	4.6 Close Project or Phase
5.Project Scope Management		5.1 Plan Scope Management 5.2 Collect Requirements 5.3 Define Scope 5.4 Create WBS		5.5 Validate Scope 5.6 Control Scope	
6.Project Time Management		6.1 Plan Schedule Management 6.2 Define Activities 6.3 Sequence Activities 6.4 Estimate Activity Resources 6.5 Estimate Activity Durations 6.6 Develop Schedule		6.7 Control Schedule	
7.Project Cost Management		7.1 Plan Cost Management 7.2 Estimate Costs 7.3 Determine Budget		7.4 Control Costs	
8.Project Quality Management		8.1 Plan Quality Management	8.2 Perform Quality Assurance	8.3 Control Quality	
9.Project Human Resource Management		9.1 Plan Human Resource Management	9.2 Acquire Project Team 9.3 Develop Project Team 9.4 Manage Project Team		
10.Project Communications Management		10.1 Plan Communications Management	10.2 Manage Communications	10.3 Control Communications	
11.Project Risk Management		11.1 Plan Risk Management 11.2 Identify Risk 11.3 Perform Qualitative Risk Analysis 11.4 Perform Quantitative Risk Analysis 11.5 Plan Risk Responses		11.6 Control Risks	
12.Project Procurement Management		12.1 Plan Procurement Management	12.2 Conduct Procurements	12.3 Control Procurements	12.4 Close Procurements
13.Project Stakeholder Management	13.1 Identify Stakeholders	13.2 Plan Stakeholder Management	13.3 Manage Stakeholder Engagement	13.4 Control Stakeholder Engagement	

Figure 104 - Project Management Groups and Knowledge Area Mapping

Anybody passing their PMI PMP certification will know the above. Many PMO organizations are staffed with PMI PMP professionals, hence the project delivery methodology of many PMO organizations follow some variation of the above.

Anybody who has been around various Enterprise PMO departments knows that sometimes companies choose to stick closely to the PMI PMBOK© standard and sometimes companies modify the PMI PMBOK© approach. The following example shows a variation:

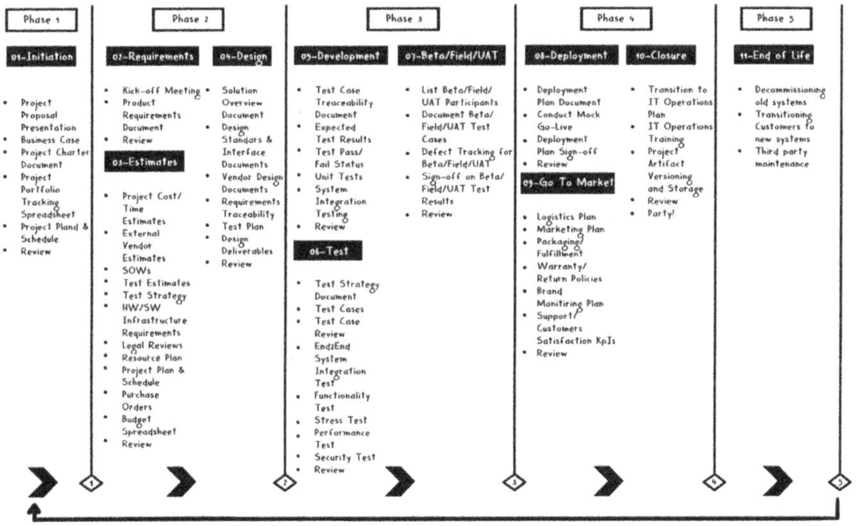

Figure 105 - Sample PMBOK Inspired Phase Gate Process

PMI PMBOK© (and its variations) is a traditional "command & control," top-down, methodology. This is by design, as it represents a generic project delivery approach, broken down into phases / process groups, controlled by phase / stage gates. Within medium and large enterprises, and IT, this is pretty much the standard process. Without getting into more details, let me also mention that ITIL® plays a similar role as PMI PMBOK©, just that it deals with more of a service-oriented view of the world. Regardless of how closely the respective process follows the standards, PMO organizations usually all provide a set of templates that can be used to navigate this maze.

Which brings us back to the Agile / Scrum Adoption Pitfall: Process Overlay.

Issues arise and become problematic when you try to marry this "command & control," top-down, approach with Agile / Scrum. What are we to do?

Agile / Scrum, being very delivery focused, plays mainly in PMBOK's **executing process group** space, but bleeds into the adjacent **planning process group** and the **monitoring and controlling process group** as follows:

Knowledge Areas	Project Management Process Group				
	Initiating Process Group	Planning Process Group	Executting Process Group	Monitoring and Controlling Process Group	Closing Process Group
4.Project Integration Management	4.1 Develop Project Charter	4.2 Develop Project Management Plan	4.3 Direct and Manage Project Work	4.4 Monitor and Control Project Work 4.5 Perform Integrated Change Control	4.6 Close Project or Phase
5.Project Scope Management		5.1 Plan Scope Management 5.2 Collect Requirements 5.3 Define Scope 5.4 Create WBS		5.5 Validate Scope 5.6 Control Scope	
6.Project Time Management		6.1 Plan Schedule Management 6.2 Define Activities 6.3 Sequence Activities 6.4 Estimate Activity Resources 6.5 Estimate Activity Durations 6.6 Develop Schedule	Agile Scrum	6.7 Control Schedule	
7.Project Cost Management		7.1 Plan Cost Management 7.2 Estimate Costs 7.3 Determine Budget		7.4 Control Costs	
8.Project Quality Management		8.1 Plan Quality Management	8.2 Perform Quality Assurance	8.3 Control Quality	
9.Project Human Resource Management		9.1 Plan Human Resource Management	9.2 Acquire Project Team 9.3 Develop Project Team 9.4 Manage Project Team		
10.Project Communications Management		10.1 Plan Communications Management	10.2 Manage Communications	10.3 Control Communications	
11.Project Risk Management		11.1 Plan Risk Management 11.2 Identify Risk 11.3 Perform Qualitative Risk Analysis 11.4 Perform Quantitative Risk Analysis 11.5 Plan Risk Responses		11.6 Control Risks	
12.Project Procurement Management		12.1 Plan Procurement Management	12.2 Conduct Procurements	12.3 Control Procurements	12.4 Close Procurements
13.Project Stakeholder Management	13.1 Identify Stakeholders	13.2 Plan Stakeholder Management	13.3 Manage Stakeholder Engagement	13.4 Control Stakeholder Engagement	

Table 16 – Agile / Scrum Focus within the Project Management Groups and Knowledge Area Mapping

The most common problem that Agile teams face working within a traditional PMBOK© based process is "double work Agile" – meaning you as the project participant have to produce all the deliverables that are required by the old process ***and*** deliver everything that Agile / Scrum demands of you – resulting in **process overlay**.

Usually the result is dissatisfaction all around:

- The Agile / Scrum process was not followed as intended, making the team less productive than it could have been – frustrating the team, the product owner, and stakeholders.
- The PMBOK© based process was not followed as intended (because the Agile / Scrum team members were too busy trying to work in an Agile / Scrum way) – frustrating existing management, which relies on the old process artifacts to measure progress / cost / success / failure.
- Everybody seems unhappy, and the new Agile / Scrum process is dismissed as "another flash in the pan."

When faced with this kind of situation, it is recommended to consider the following:

1. How do you make your Agile / Scrum team most effective? What are they good at (do those!)? What aren't they good at (seek support from others on areas that Scrum participants are really bad at!)? For example, Scrum masters ***are not*** project managers.

2. What are the things you will not be able to change? For example, some organizations use time sheets to do detailed work breakdown analysis and cost control of vendors – if that is the case, do not fight it; learn how to work within the system to minimize impact on the Scrum team while providing the information necessary to support the organization.
3. Review the project management process groups and determine what falls clearly outside of your Agile / Scrum team's responsibilities. Exactly what is your Agile / Scrum team responsible for versus where do you need support from PMO folks? For example, a project charter should be done outside of the Agile / Scrum team, as should the identification of the stakeholders. Closing the project or phase, as well as closing any procurements, should be outside of your Agile / Scrum team.
4. Next, negotiate with the management team, the stakeholders, and whoever is responsible for the PMBOK© based process the areas that overlap (see above). For example, **collect requirements** will be covered in the **user story / backlog grooming / refinement** process – make that official and get the PMBOK© item off the list. Be explicit about who will do what. Be explicit about the requirements documentation standard – user stories, not arcane 124-page-long functional specifications. Some areas that are listed under PMBOK© will simply go away in Agile / Scrum. Explain to the PMO organization how Agile / Scrum will replace existing processes. If deliverables, such as reports, are derived from the process

groups, understand what they are and how Agile / Scrum can support those / replace them. For example, it is not uncommon to see management reports that took great manual effort being replaced by data dumps of data points from the continuous integration server on an Agile project. Instead of reporting on the weekly build, you are able to show you build 56 times a day, out of which 43 builds were successful, the number of defects found vs. fixed, and showing the build stabilization trends over time. Which report do you think is a bigger hit with executives?

5. For each of the areas that are covered in the core **executing process group** (i.e., the group that is most directly impacted by the Agile / Scrum team / effort), show the cold, hard numbers. For example, under **perform quality assurance**, show the number of continuous integration builds (success / fail), the number of unit tests, unit test coverage, automated functional tests, number of environment supported / tested, etc.
6. Tailor and adapt as needed. Rinse and repeat.

There you have it. It is not pretty, but we don't live in a perfect world. Fact is that many Agile / Scrum projects have to live within existing process frameworks, so it's up to the Agile / Scrum team and sponsor to negotiate how the new Agile / Scrum adoption effort fits into the enterprise project management picture. It is acknowledged that it is not ideal, but it is also generally accepted that this is the case with 80% of all projects.

CHAPTER 29
LACK OF ENGINEERING BEST PRACTICES

Engineering best practices are the baseline that companies need to establish in order to improve upon the development cycle with any methodology – Agile or not. As long as a company has issues with these best practices, they are doomed to fail for any serious process improvement effort.

Many companies seem dead set on entering into an Agile / Scrum adoption effort but struggle with the basics of software development.

Over the last 50 years, core engineering best practices have developed that form the basis for any successful software development effort. Please keep in mind that although engineering best practices form the basis for success, they do not necessarily guarantee it. Many companies follow all the best processes but miss the boat when it comes to delivering the right solution to market.

Before we go further, let us clarify what makes up a "best practice" to begin with. According to ***Merriam-Webster***[50], a best practice is defined as "*a procedure that has been shown by research and experience to produce optimal results and that is established or proposed as a standard suitable for widespread adoption.*"

An analogy to clarify the meaning of a best practice is learning a craft or becoming really good at a sport – you need to know the basics before you can become really good. Best practices are the building blocks that allow you to be successful. If you lack the basics, you are doomed.

For example, as a carpenter, you need to know about what kind of wood to select, what tools to use, how to use those tools proficiently, how to prepare a work site, how to measure and estimate things, and finally how to construct things in order to be successful.

In soccer, you need to build up your core stamina so you last 90 minutes, you need to master ball control, you need to understand your specific position on the team, you need to know basic soccer offensive and defensive strategies, etc.

Unless you know these basics, these kinds of best practices, you will not be successful, no matter what your talent. Best practices have

---

[50] http://www.merriam-webster.com/dictionary/best%20practice

to be internalized – meaning they must happen automatically without much conscious effort, repeatedly – in order to be effective.

This last point is important. **Best practices have to be internalized by the team and the company in order to be effective**:

- Michael Jordan was able to make amazing shots 2 seconds before the buzzer rang because all the techniques that came with him jumping off the floor, raising his arms, directing the ball, focusing in on the basket, all those effortless skills were internalized.
- Good soccer teams move up and down the field in unison as the ball moves along the field because they internalized that behavior.
- Good software teams make sure their source code for their emergency fix that will solve the world-wide system outage is peer reviewed, checked-in, and fully tested.

Internalized behavior basically means you are in the "zone" – you know what do, and you do it without conscious decision. Not to belabor the point, but until you and your team or company have internalized best practices, many Agile / Scrum adoption efforts (or any other significant process improvement effort for that matter) will fail because the basics are not in place. You cannot build a house on a weak foundation.

Here is a short of list of "must have" engineering best practices:

- **Develop iteratively** – For most modern software development projects, some kind of spiral-based methodology is used, rather than a waterfall process. There are myriad choices out there, from older legacy models such as the Rational Unified Process (RUP), to the latest Agile / Scrum, and Extreme Programming (XP) frameworks. Having a process is better than not having one at all, no matter what process you are following. In general, it is less important what process is used than how well it is executed.
- **Establish change / requirements / configuration management** – If you want to build good software, you need to express what you want to build, you need to track it, and you need to manage it by versioning it. Agile / Scrum focuses on user stories for expressing requirements; other methodologies go into various levels of depth and details using other formats (RUP utilizes use cases and UML, for example). Configuration management involves knowing the state of all artifacts that make up your system or project, managing the state of those artifacts, and releasing distinct versions of a system. You need to know what you build, with what version of the code, on what version of the environment.
- **Use component architectures and design patterns** – Depending on your technology stack, you must pick and choose the appropriate architecture for your application. Component technologies, web services, and commonly used design patterns all provide best practice guidance for your application –

useful knowledge of what works, what does not work, and why. Most important, these standards and patterns help you to not reinvent the wheel but to stand on the shoulders of others before you.

- **Use source control** – Disregarding the often emotionally charged discussion about tool preferences, from this author's perspective, it does not really matter if a team uses Git, SVN, CVS, Team Foundation Server, or any other source control system – as long as you use it effectively. All of these systems have advantages and disadvantages, but in the end they are very much like choosing an Uber to get you to the airport – some are nice, some are big, some are small, but they all get you there.
- **Implement unit testing** – A best practice for constructing code includes the daily build and smoke test. Martin Fowler[51] goes further and proposes continuous integration that also integrates the concept of unit tests and self-testing code. Kent Beck[52] supposedly said that "daily builds are for wimps," meaning only continuous integration would provide the best benefits from a coding perspective. Ultimately, the frequency needs to be decided by the team, but the more frequent, the better – at least based on the author's experience. Even though continuous integration and unit tests have gained popularity through XP, you can use these best practices on all types of

[51] http://martinfowler.com/

[52] http://www.amazon.com/Kent-Beck/e/B000APC0EY

projects. Standard frameworks to automate builds and testing, such as Jenkins, Hudson, Ant, and xUnit, are freely available.

- **Conduct peer reviews, code reviews, and inspections** – Make sure you know what is in your source code. It is important to review other people's work. Experience has shown that problems are eliminated earlier this way, and reviews are as effective as or even more effective than testing. Reviews and inspections also are effective as a learning / teaching tool for your peers.
- **Perform testing** – Testing is not an afterthought or an opportunity to cut back when the schedule gets tight. Disciplined testing in conjunction with extensive unit testing and reviews guarantees that software works. It is an integral part of software development that needs to be planned and executed. Platform proliferation (various operating system versions, various browsers, desktop vs. mobile platforms, etc.) has made testing critical to any project's success. Testing still requires a good amount of manual effort, but compatibility testing across various platform combinations can be effectively automated using tools such as Selenium.
- **Automatically measure the process** – How many developers are committing code every day? How many lines of code or class files? How many unit tests are executed per continuous integration run per day? As the project is coded and tested, the defect arrival and fix rate can help measure the maturity of the code. It is important to use a defect tracking system that is

linked to the source control management system and the automated build system. By using defect tracking, it is possible to gauge when a project is ready to release by correlating information about the volatility of the code base to the defects found / fixed based on the set of tests executed. This kind of measurement has to occur automatically. Especially with continuous integration and automated platform tests, generating metrics needs to be automated due to the high frequency and large data volume that gets generated.

- **Use smart tools** – There are more and more effective tools that help software development teams address all kinds of problems, from coding standards, to security scans. Utilize as many of the **static**[53] and **dynamic**[54] code analysis tools as you can in as much of an automated way as you can to make your life easier.

That sums it up. Admittedly, for some teams or organizations this list of best practices might seem intimidating, but unless you invest in making these best practices work, you will not see meaningful payback to any process improvement effort or any Agile / Scrum adoption effort. At best you might experience partial success in some areas, but the larger benefits of key process improvements or Agile / Scrum adoption will be denied – because you cannot build a house on a weak foundation!

---

[53] http://en.wikipedia.org/wiki/List_of_tools_for_static_code_analysis
[54] http://en.wikipedia.org/wiki/Dynamic_program_analysis

Chapter 30
Ignoring DevOps

Agile / Scrum is all about cutting down on long communication paths between the business / product owners and the development team, essentially delivering software solutions that meet the business and the product owner's expectations. However, Agile / Scrum does not by itself address the operational aspects of running the software solution in a real world environment. That's where DevOps comes into play.

While Agile / Scrum is very much focused on improving the productivity of the development team, DevOps is focused on automating the entire delivery value chain of software and providing automated feedback loops, taking a full life cycle view of software development, from business planning through operations.

Some key "best practice" processes that underlie the DevOps are:

- **Developing in "small batches,"** i.e., building increments of a minimally viable product – A key concept of Agile / Scrum development is to develop a minimally viable product in small increments. This has multiple advantages: a) the planning horizon stays manageable and rooted in the "now"; b) it builds confidence in the team; c) it proves to the product owner that delivery is feasible; d) small increments are better than "big bang" changes from a user's perspective, etc. Think about your car – from model year to model year, changes are incremental. The same concept applies here
- **Test-driven development**, or at a minimum the implementation of a disciplined unit testing approach – Test-driven development (TDD) is a software development process that relies on the repetition of a very short development cycle where requirements are turned into very specific test cases, then the software is improved to pass the new tests. This is opposed to software development that allows software to be added that has not proven to meet requirements. The key point here is that each requirement is developed based on test criteria. Whether your project follows this strict TDD approach, or if you simply ensure that your software has corresponding automated unit tests for every software change, is beside the point as long as you implement tests for every change. It is hard to stress how much this approach improves quality, system stability, and maintainability – it is really the basis of building a

quality product from the ground up.

- **Instrumentation**, across the product or solution, including infrastructure – Instrumentation at all levels is critical for keeping a handle on your software, your infrastructure, and ultimately your customer satisfaction. Every system built is different, and every specific domain has its own challenges, but simply put, you should be able to tell at any given time:
    1. How many users are on your system
    2. What their activity level is
    3. The number of transactions that succeeded / failed
    4. The utilization numbers for various server-side health indicators like memory utilization, database connections, uptime, disk space utilization, etc.
- **Continuous integration** – Continuous integration (CI) is a development practice that requires developers to integrate code into a shared repository several times a day. Each check-in is then verified by an automated build, allowing teams to detect problems early. CI allows for 3,000-mile-high monitoring of the development team – especially important in today's distributed development model – such as understanding who checks in code how frequently, who breaks the build, how many tests fail, how many tests are added, etc.
- **Automated builds** – Build automation is the process of automating the creation of a software build and the associated processes, including: compiling computer source code into binary code, packaging binary code, and running automated tests.

Automated builds are usually a separate task / activity from CI, because automated builds focus on packaging the software solution into something consumable for a target test environment (for example by automatically deploying to a set of virtual machines for testing purposes), whereas CI only tries to accomplish its tasks within the CI environment.

- **Automated testing** – Automated testing tools are capable of executing tests, reporting outcomes, and comparing results with earlier test runs. Tests carried out with these tools can be run repeatedly, at any time of day. The method or process being used to implement automation is called a test automation framework.
- **Continuous testing** – Continuous testing is the process of executing automated tests as part of the software delivery pipeline to obtain immediate feedback on the business risks associated with a software release candidate.
- **Application release automation** – Application release automation refers to the process of packaging and deploying an application or update of an application from development, across various environments, and ultimately to production.
- **Test-driven deployment** – Test-driven deployment refers to the approach of automatically validating, via automated test scripts, that the deployment of a build or release to the target environment was successful. This can range from simple smoke tests on the target environment to full, detailed, end-to-end tests that validate that the deployed functionality is

working as expected.

- **Continuous delivery** – Continuous delivery (CD) is an approach in which teams produce software in short cycles, ensuring that the software can be reliably released at any time. It aims at building, testing, and releasing software faster and more frequently. The approach helps reduce the cost, time, and risk of delivering changes by allowing for more incremental updates to applications in production. A straightforward and repeatable deployment process is important for continuous delivery.
- **Canary release rollouts** – Canary releases are designed to reduce the risk of introducing a new software version in production by slowly rolling out the change to a small subset of users before rolling it out to the entire infrastructure and making it available to everybody. A good example of this is how Android app publishers can configure their app for Google Play Store Beta distribution accordingly, but the approach can be done regardless of the technology used. As a matter of fact, the author used this very approach some 25 years ago when shipping an accounting software package to some 30,000 users. We would ship the new software (on CDs) to 1,000 users and wait for feedback; with no complaints; we would roll out the next 2,000 copies, wait again; then release to the remaining user base. Another approach used today is the so called Blue/Green/Deployment approach (sometimes called A/B/Deployment), where you have two similar production en-

vironments, one with the current version (Blue/A), one with the future version (Green/B), and you switch users from one environment to the other without their knowledge. Again, regardless of what approach is used, the purpose is to make the deployment of software release transparent to the end users and minimize risk to the company.

- **Develop and test against production-like systems** – This concept stresses the importance of developing and testing on production-like systems, down to the specific operating system patch level and supported libraries. Also, if you deal with localization / internationalization projects, make sure to run your application / app on the localized system just as your end users will.
- **Utilize automated feedback loops** – The last obvious one to mention is automated feedback loops – and this is closely related to the aforementioned **instrumentation**. Let me stress the need for automation here again – this cannot be a manual effort; it has to be automated. With all the approaches / processes mentioned above, many of them automated, there is simply no way for a human to keep up. Feedback loops have to run constantly, update dashboards, and alert the right folks at critical junctures. For example, if a system runs "hot" and runs out of memory, an alert has to go out to your DevOps engineer to review what is going on; that DevOps engineer might in turn pull in developers to see what is wrong. The automation might even take down this server and bring another server

online. Unless this is automated, there is no way the feedback loop will be effective – it will be too slow, and in the above example, your hot running system will crash.

Please note that the above listings are mainly DevOps processes. Besides the process perspective, you will need to look at the technology, people, and culture aspects of DevOps in order to make it successful. You might have all the right technologies and process capabilities, but your people still look at weekly build as the way to go; or the other way around, you got the right people but you did not acquire the right technology in order to cut wait times out of the value stream.

All these things – processes, technology, people, and culture – have to come together in order to make DevOps work effectively in an Agile / Scrum context. But DevOps by itself is not specific to Agile / Scrum – you can make it work with all kinds of development methodologies / frameworks.

DevOps is important in the context of Agile / Scrum adoption success because a highly successful Agile / Scrum team might still be perceived to have failed if the operational side of running their software solution is problematic. In that sense, DevOps truly focuses on successful delivery of the entire software solution, end to end, not just the binaries that are the output of the Agile / Scrum team effort.

About the Author

Brian is passionately committed to helping companies make the transition to Agile / Scrum and increased team productivity.

His most recent professional focus has been primarily on software development consulting within the Predictive Analytics / Machine Learning / Big Data space, as well as Mobile App Development on iOS and Android.

In the past Brian worked within the commercial software product space in various functional capacities – covering diverse problem areas, from operating system and compiler development, to enterprise financial systems and mobile apps. He also has substantial experience with Ecommerce systems transacting in excess of $3B.

Brian's background is rooted in the operating system (AIX) and the object-oriented technology community. He started as a software engineer porting old proprietary COBOL code to modern UNIX and C architectures, led quality assurance teams for UNIX kernel development efforts, and managed several functional quality and test organizations for commercial Smalltalk compiler teams.

Based on his early exposure to Agile thinking within the Smalltalk community, Brian was an early adopter of Agile practices, including the early application of Kent Beck's now famous Smalltalk unit testing framework during the VisualStudio and VisualWorks development cycles – which is now known as the xUnit framework, supporting a multitude of programming languages.

During the course of his career, Brian has held many roles, including successful stints as a Scrum Product Owner, Scrum Master, Agile Coach, and member of development teams.

In addition, he has held numerous functional management roles: VP of Professional Services, Senior Program Manager, Director of Quality Assurance, and Director of Quality Engineering. He was also part of an early adopter team implementing real-time, in memory, predictive analytics for credit and debit card fraud detection.

His multifaceted background gives Brian the ability to understand (and explain) Agile / Scrum and its implications equally well from multiple perspectives: from the development team to the executive board.

INDEX